basic

ECOLOGY

"THIS it is that makes the Amusement
of Life—to a speculative Mind—I go among
the Fields and catch a glimpse of a Stoat
or a fieldmouse peeping out of the withered
grass—the creature hath a purpose, and its
eyes are bright with it. I go amongst the
buildings of a city and I see a man hurrying
along—to what? the creature has a purpose
and his eyes are bright with it."

KEATS, 1819

basic
ECOLOGY

by

Ralph Buchsbaum, Ph.D.
Professor Emeritus of Biology
University of Pittsburgh

and

Mildred Buchsbaum, M.S.

Pacific Grove, California

Twenty-third printing, 1990

Distributed by

THE BOXWOOD PRESS
183 Ocean View Boulevard
Pacific Grove, California 93950

Phone: 408–375-9110.

Library of Congress Card
No. 57-28684.

Standard Book Number: 912086-05-1

Printed in U.S.A.

PREFACE

THE WORD ECOLOGY is beginning to appear regularly in newspapers and magazines, but there is still an unfortunate lag between the use of the word and the general diffusion of ecological understanding.

Ecology has not lent itself well to the brief treatment accorded it in a short chapter of a general biology textbook. The insights of ecology come to the beginning student only when he has had pointed out to him, with special emphasis and in vivid detail, how his own experience as a member of a human community and his miscellaneous bits of information about the natural world as he knows it from garden or woods, can be integrated into meaningful ideas. The ecological "niche" as applied to either human or natural communities the world over, ecological succession as a unifying explanation of the patchwork of fields and ponds, forests and dunes that we see around us—these are concepts that deserve a place in the education of college students towards a mature citizenship.

Our world is one in which ever-expanding populations, depletion of natural resources, water and air pollution, wise land use, the building of huge and not necessarily useful dams, dust storms, recurring floods, management of game and recreational areas, are all matters on which the citizen must have informed opinions. It is difficult to see where the foundations of such opinions are to be built if not at the first-year college level and on the structure afforded by an understanding of basic ecological principles illustrated in the familiar context of the student's own experience.

The esthetic values that come from knowing the ecological and historical relationships between a pond and the surrounding woods, between a sandy beach and the forest in the dunes that rise above it, between cultivated fields and the grassland or forest that they displaced, are like the pleasurable insights that come with understanding the literary and historical background of western civilization. The student cannot appreciate the ecological setting in which he lives without knowing of the historical, economic, and political events that led to the pollution of the rivers of Pennsylvania, to the cutting of the forests of Ohio, to the breaking of the sod of Montana. Nor can he understand the historical and political growth of his country without an ecological understanding of why the cotton economy of the South began to exhaust southeastern soils even before the Revolution while the prairie sod of Iowa is among the least exhaustible soils in the world, or how the Gulf Stream has influenced the literary output of England.

Twenty years ago the senior author of this book could find no reading materials suitable for supplementing the lectures and discussions of the general ecology section of a course at the University of Chicago that was designed to provide students with the contributions that could be made by biologists to a broad general education. As a temporary expedient he prepared a small booklet called "Readings in Ecology."

Since that time many ecology textbooks have been published. For the beginning professional biologist or ecologist at the upper college level there are now excellent texts of general and of plant ecology by Clarke, Odum, Woodbury, Daubenmire, and Oosting. The graduate student has available the comprehensive "Principles of Animal Ecology" by the "Chicago Group" of ecologists. However, at the first-year level, and for the general education of students without special interests in biology, there is still no textbook of general ecology. The authors have been preparing one for some years, but it is not ready. The old "Readings in Ecology" has been allowed to go out of print. To fill the present need, and to try out some of the materials being prepared for the somewhat larger book, which will be different in some important ways, in both content and emphasis, we have prepared BASIC ECOLOGY.

Whether the selection of content in so small a book has been wise, and whether the ecological views are accurately and clearly expressed is of course the responsibility of the authors.

The views themselves have certainly been molded by many years of stimulating association with our teachers and friends, the late W. C. Allee, A. E. Emerson, Thomas Park, Karl P. Schmidt, Orlando Park, and with their many students of ecology. Most of the drawings were executed by Bernard Garnett; a few by Patricia O'Brien.

THE AUTHORS

January 1, 1957   viii

CONTENTS

Chapter   Page

1   1   WHAT IS ECOLOGY?

2   20   THE PHYSICAL ENVIRONMENT

3   39   THE LIVING ENVIRONMENT

4   58   THE COMMUNITY

5   83   PERIODIC CHANGES IN COMMUNITIES

6   98   ECOLOGICAL SUCCESSION

7   128   THE DISTRIBUTION OF PLANTS AND ANIMALS

8   141   BIOMES OF THE WORLD

9   178   CLIMATIC GRADIENTS IN PLANTS AND ANIMALS

       185   BIBLIOGRAPHY
       189   INDEX

Chapter 1

WHAT IS ECOLOGY?

"A PERSON should not shoot a bird resting on his own head" is a saying of one of the Bantu tribes of South-West Africa, and it is not intended to be a guide to the hunter. Any Bantu who can hunt hardly needs such advice. It is a legal maxim, and advises anyone engaged in a legal dispute not to testify against a relative lest he harm himself. Since one's interdependence with some other person is not always obvious, the proverb is put into terms of a direct physical connection that anyone can understand.

Though we cannot carry around on our heads, where they would serve as vivid reminders, the soil in which we grow our food, or the birds that eat our crop pests, or any of the other myriads of things in our environment on which we depend directly, or indirectly, for our survival—we should carry some understanding of all this in our heads. Those people who do are said to have an "ecological viewpoint."

Such a viewpoint is neither restricted to scientists, nor even widespread among all the different kinds of specialized scientists. It requires no great fund of detailed information, or wide experience with experimentation or scientific method. It requires only enough appreciation of the complexity and interdependence or wholeness of the living world to inspire a cautious approach to intervention in so complex a mechanism. Anyone who knows enough not to try to repair his expensive wrist watch by poking around in the works with a hairpin, but who instead goes for help to an expert repairman, will understand why the ecological viewpoint counsels careful study before doing such a thing as replacing with cropland a forest or a prairie that has been successfully established for thousands of years.

The ecological viewpoint as applied to man's problems is not a special kind of view. It is merely the application to practical eco-

logical problems of the same views we consider reasonable in other
fields.

If you were to go to the doctor complaining of shortness of
breath and if you weighed 350 pounds, the doctor would probably
advise you to reduce your weight, not to exercise to enlarge your
heart. Yet the same doctor, relaxing at home with a newspaper,
and reading that world population growth is increasing at a greater
rate than any possible increase in food production which we can
project for the near future, thinks that perhaps we should solve
this problem by putting more effort into learning how to harvest
algae from the sea.

If water were dripping through the ceiling onto the floor, you
would not call a plumber to install a special drain in the floor.
You would instead hire a roofer to see why water was coming
through the roof and through the ceiling. If he found the gutters
intact, he would examine the roof. He would try to stop the trouble
at the source. Yet when we have really disastrous leaks, like
floods that do millions of dollars worth of damage and take many
lives, we feel we are advanced in our thinking if we merely agree
with some engineers that we ought to build higher levees and more
dams. The fact that many dams we have built in the past are
already silted up, some of them within seven to ten years after
construction, seems only to suggest to us the building of new dams
above the old ones. The money spent on a single dam is much better invested in an adequate program of reforestation and ecological
control upstream, where floods begin. Such measures become increasingly effective as time goes on. And forestland can steadily
provide lumber if good forestry practices are followed.

A farmer who found a screw on the ground next to his tractor
might not know exactly where to replace it, but he would not just
throw it away. He would understand that in a complex machine
even one small part may be essential for the functioning of the
whole machine. Yet the same farmer, once having seen an owl
capture a small bird concludes that all owls are a menace to his
chickens and thereafter kills all owls—not considering that an owl
might be an integral, and possibly very useful, part of the environmental complex of his farm. He does not ask himself what the owl
was doing on all the days of the year when he was not making off
with a bird. One Pennsylvania farmer with an ecological viewpoint
did ask himself this question, and he went on to collect 203 pellets

of indigestible residue dropped by barn owls on his farm. He took the pellets to Carnegie Museum, where examination disclosed that the owls had eaten: 429 meadow mice; 4 lemming mice; 1 pine mouse; 12 white-footed deer mice; 18 jumping mice; 21 star-nosed moles; 1 Brewer's mole; 95 large short-tailed shrews; 1 least short-tailed shrew; 1 squirrel; 5 cottontail rabbits; 23 unidentified mice; and 5 small birds. Now this farmer offers special protection to a pair of barn owls that nest atop his silo. He considers them an indispensable asset to his farm.

Barn owl with field vole. (Photo by Eric Hosking)

THE SCIENCE OF ECOLOGY

The organized body of knowledge which deals with the interrelationships between living organisms and their environment is a relatively new science, which we call ECOLOGY. The term is derived from two Greek words meaning a "study of the home," and it has been in use only since the 19th century. But the observation of plants and animals in their natural "homes" has been going on during all of the million and more years that men have sought them as food or for clothing or avoided them as enemies. Knowledge of just where and when to gather wild fruits and grains, which tides uncover the most shellfish, how best to approach a fox with a club, and exactly how deep to plant seeds—such things are well worth knowing and then passing on to one's sons or tribesmen.

The steady accumulation of folk knowledge served as the basis

4 BASIC ECOLOGY

for the more systematic collection of facts about nature that eventually came to be called <u>Natural History.</u>

In 1735 the great Swedish botanist, Linnaeus, published his
"Systema Naturae," the first good set of rules for describing and
naming plants and animals, and the one we still use today. This
set off a new era of "systematics," the scientific description and
classification of organisms. The systematists became caught up
in a frenzy of collecting and classifying that carried them to all
parts of the world. It is difficult for us now to recapture the almost
limitless expectations of men who lived at a time when ships were
scouring the southern oceans in search of whole new continents.
Australia had just been discovered, and New Zealand, and many
lesser but exotic islands. And the fact that the "great southern
continent" was never found did not make the expectation of its dis-
covery any less exciting to naturalists and collectors. They spent
much time poring over maps to discover where they could go to
find new species.

Great museums were opened to exhibit the captures of explorers
and the finds of collectors. It was this period that contributed the
caricature of the bespectacled professor chasing butterflies with
a net, still a popular notion of a biologist.

To these early naturalists and systematists we owe a large
part of our knowledge of the classification, the distribution, and
the habits of plants and animals. With the rise of modern experimental biology, descriptive natural history has given way to the
more functional and more quantitative study of plants and animals
in their environments, which we call ecology.

The beginnings of ecology go back into the past farther than
those of any other science. But as a body of principles it has lag-
ged behind other sciences; and necessarily so because of its com-
prehensiveness. Including within its theoretical scope all physical
and all biological phenomena, ecology has had to await the development of many other sciences before it could advance much beyond the descriptive stage.

In the physical sciences, where problems deal with a very
limited number of variable elements, great exactness and mathematical treatment were achieved long ago. In ecology, where all
problems are relatively much more complex, involving many vari-

able components of the environment, exact measurement and mathematical treatment are relatively recent and just beginning to yield important conclusions.

THE ROLE OF THE ECOLOGIST

The ecologist continues to add to the accumulation of facts that tell what things there are and what happens in the natural world. But he sees it as his primary interest to organize the chaotic collections of facts, old and new, into generalizations (principles and laws) which help to explain how these events take place where and when they do, and to predict what will happen under a given set of conditions.

Predictability, the real test of a scientific generalization, is infinitely more difficult in ecology than in less complex sciences. It is not of the same order as that in chemistry, nor can it even be compared to that in human physiology, which limits itself to the internal mechanisms of a single living species. Nevertheless, ecology as applied to man's problems is already in a position to make the rough predictions man needs to act intelligently. For our biggest ecological problems we do not need the accuracy used in timing an eclipse or estimating the yield of a chemical reaction. It does not take an exact science to predict agricultural poverty if we continue farming practices that in twenty years remove six inches of indispensable topsoil that took 3600 years to form! An ecologist can look at a forested mountainside in a particular region and easily predict what will happen to the farms in the valley below when all the trees on the mountainside have been cut. If the flood damage turns out to be only $385,000 worth of ruined cropland and buildings instead of the $500,000 worth of damage he predicted, the ruined farmers will not quibble about the lack of exactness. They may learn instead to take such predictions more seriously.

None of us is really unaware of the problems man creates in rearranging his environment to suit his wishes. The newspapers are filled with news of parched crops in the Southwest, calamitous floods in the Ohio River Valley, the return of the destructive Mediterranean fruit fly to the orchards of Florida, agricultural poverty in New England, and even "humorous" stories like the one about the large city in Texas where drinking water was selling for fifty cents a gallon while oil was seven cents a gallon. For those of us who only read about such matters but are not visibly

affected by them at the moment, it is only natural to be optimistic
and to reflect that nowhere else in the world are there so many
people as satisfied with life as in our own country, in spite of our
troubles.

Two hundred millions of people are so well fed that our major
problem in agricultural economics is how to dispose of vast surpluses of food. At the same time our public health officials warn
that over-eating has become one of our most serious health hazards.
People who grew up in farmhouses that were heated with wood and
lighted with oil lamps, now turn on radiant heating or electric light-
ing at the flick of a switch. We can watch the events of a whole
world come to us on our television screens, and alter our summer
climate with air-conditioning. One disease after another is rapidly
being brought under control. And the future promises more and
more of everything for more and more people, provided they can
learn to get along with one another. Hunger, want, and disease are
becoming a thing of the past, say those who hail our constantly expanding population as a "built-in" guarantee of a steady increase
in consumer demand for goods and services.

Then suddenly we become caught in some personal crisis, perhaps nothing more serious than a traffic jam. As we inch along, we
begin to feel with the man in the "Punch" cartoon. Caught in a never-
ending line of cars, he keeps grumbling about how much worse the
traffic is than last year. His wife glares at him and says, "Just be
thankful it's not next year!" Whether we shall be thankful for next
year, or merely thankful that it's not the year after next, we
shall in time know.

The American continent, home of the most widely distributed
wealth in the history of large civilizations, has "aged" faster than
any comparable land area in the world. The people most informed
about food supplies around the world tell us that our surplus of food
is a special and only temporary phenomenon in the United States
and in a few other areas of the world, that two-thirds of the peoples
of the world are badly under-nourished, and that there is no clear
prospect that food production will keep pace with the increase in
population everywhere, much less provide more adequately than
at present.

Population experts describe current rates of human population
increase as a "population explosion" and our period in history as
"a human plague. " They point out that public health measures have

reduced human mortality to record lows, while at the same time birth rates continue at a new high level. They calculate that if the current rate of population increase had been in effect regularly for the last 5300 years (a mere fraction of man's history) the earth would now be solidly covered with human bodies to a depth of several miles! Lucky for us that the expansiveness we look on as our privilege, started only in the 19th century. It will have to be checked if man is not to meet the same fate as all other species which go through temporary increases we call plagues.

What of the energy sources that support our industrial civilization? The manager of research for one of our largest oil companies says: "the day of oil shortages is now almost upon us and . . . the peak of the fossil fuel [coal, oil, and gas] era will come before the end of this century. The inventive genius of our own country, in particular, will have to turn away a little from the present overemphasis on entertainment, convenience, comfort, thrill, and appearance to the more bread-and-butter subject of the raw need for energy." An unpleasant phrase, "the raw need for energy," so we turn our attention from the oil experts to those atomic scientists who promise us a life of ease as soon as they harness atomic energy. They are likely to make good their promises, but nothing is perfect, we sadly realize. The by-products of atomic fission are highly destructive to protoplasm, and in unique ways which present problems that never were posed by the other hazards man has introduced into his environment by his technology. The dangers are at present underestimated, rather than exaggerated, in the opinion of most biologists.

At the present time we still have two-thirds of the topsoil we found in the United States, population density has not yet outrun food supplies, fuel shortages will not begin to be really felt for some time, and small but repeated doses of radioactivity do not generally show their insidious effects immediately (often they do not produce serious effects until twenty years or more after the damage is started).

Since human societies seldom take any effective action until a crisis is actually upon them, or until it is already too late to act effectively, it is likely that all current trends in the United States will continue for some years yet, especially since we are in the midst of an unprecedented expansion of both the population and our industrial capacity and in no mood to worry about future consequences.

The ecological viewpoint is the integrated view of man's environment, stressing the interdependence of all its parts. We shall now see what light this throws on a number of man's important ecological problems.

SOIL CONSERVATION

Any problem as old as that of the maintenance of soil fertility can hardly be approached in a young country like ours without glancing around first to see what has happened to our elders. The most fertile valley in the world, the valley of the Nile, has not lost its fertility. Its rich mineral and organic content is renewed by silt brought down by the river during the annual flooding and deposited on its flood plain. But the rich civilization it once supported is now replaced by unimaginable poverty, perhaps the worst anywhere. The population density is greater than in the most populous parts of China, and even this most marvelously rich soil will support just so many people adequately—and no more.

Of the other great river valley civilizations much of Babylonia (now Iraq) is a sandy waste; India is a country of recurring famines in which millions of people die; and China brings to mind not only great famines but also the world's most spectacular floods. These floods are mostly those of the Yellow River in the north, where the headwaters of the river descend from deforested areas and eroded plateaus that have been so denuded of vegetative cover by thousands of years of human activity that plentiful rain is no longer a blessing to the land, but an uncontrollable menace.

It should be no comfort at all to us to dismiss these as "backward" areas of the world. The Nile valley is still incredibly fertile. India from 1000 B.C. to 500 A.D. was in the same vigorous actively expanding stage we are in now and exported so much manufactured goods to other countries that the Romans worried about the steady flow of gold from Rome to India. China's troubles are to be laid more to long occupancy of the same lands by too many people rather than to any gross negligence or ignorance. The Chinese made land management,which we call soil conservation, an official policy of the government in 2700 B.C. The forests of China were the only important source of fuel in the centuries before man had learned to borrow energy from the past by using the fossil fuels. The grassy plateaus were gradually denuded by grazing animals and by an agriculture that was by no means as wasteful as ours. The agricultural experts we sent to China in the 1930's came home impressed not with the backwardness of the Chinese but with the many ways in which their agricultural methods were in advance of ours and with the ability of so many people to live for so long a time in one place without making many of our mistakes. In the rice-growing southeast of China they found incredibly high soil yields and the fertility of the soil apparently unchanged for at least the last 2000 years.

How does our own record compare with all this? Not nearly so well as we would like to think. We started at a later stage, could have been better informed, and have not even yet suffered the population pressure these peoples have had for a long time. We found "a land of unlimited forests and fertile plains." But we unwisely converted the thin forest soils to cropland and when they failed to yield well after a few years abandoned the farms or shrewdly sold them to later comers and moved westward—to start new farms on cutover forest land. A state like Ohio, for example, was nearly 95% forest-covered in 1788. It was still 54% woodland in 1853. And only a declining rate of deforestation saved it from going below 14% in 1940. It has not changed much since then.

In the common view of things the pioneers who cut the forests of Ohio, and to do so braved not only Indians but cold winters and many hardships, deserve nothing but admiration for "developing" a young country and for leaving us the rich heritage of a continent whose resources were opened to us by their work.

The ecological view is somewhat different. The pioneers must be admired for their courage and their herculean efforts to build a new world. But they were uninformed, and often too hasty if not greedy. Their greed was no worse than ours today, and like much of ours it came not out of any intent of doing harm to those who have followed but out of mistaking short-run benefits for long-term good. The removal of the forests in Ohio and in much of the Ohio River Valley has resulted in a denuded soil which simply cannot soak up the heaviest rains. Once the millions of trees, each with thousands of leaves, broke the force of pounding rain, cascading the raindrops downward from one leaf to another until they tumbled gently onto the deep, spongy humus mat of the forest floor, where drops continued to fall and to be absorbed for hours after

Plow, breaking the virgin sod of the grassland in the great plains, near Park River, North Dakota. This man's effort to carve out a niche for himself as a farmer, to support himself and his family, and to produce food for the general population, is both necessary and commendable. But this action requires more than mere labor; it requires proper attention to ecological mechanisms which, if ignored, may ruin the land. *Below,* is a typical abandoned farmstead which tells the story of years of drought and crop failures. It was farmed without an adequate understanding of what this land in this Texas climate could produce. (Photos on this page and the next by U. S. Dept. of Agriculture.)

A bad dust storm on a field which recently had been burned of Russian thistles preparatory to plowing. However, the thistles were a result of unwise land use on this farm near Culbertson, Montana. A great deal of knowledge is needed to manage the land to make it productive and to conserve it as well.

Above, the results of uncontrolled wind erosion on a farm near Culbertson, Montana. It is important for every farmer to know how to manage his farm, not only so that it serves his own interests, but so that it will not blow away and spoil his neighbor's land.

Below, a contour planting of wheat with alternate fallow and wind buffer strips is one method of protecting the land against wind erosion in this field near Power, Montana.

the rain had stopped. Now raindrops fall with force on bare or
sparsely covered soil which, through long years of erosion has
had much of its humus removed, leaving behind a hard pavement-
like surface of close-packed soil and pebbles.

Wherever rain falls on bare soil, on soil made hard by erosion
of its spongy upper layer, on soil laid bare by the teeth of too great
a concentration of cattle or sheep, on soil laid bare by fire, the
water cannot soak in readily. Much of it runs off, carrying with it
a load of the finest materials and leaving behind an increasingly un-
absorbent surface. The compacted soil now sheds 50% or more of
the rainfall, which may run off in great sheets into small gullies
and then large ravines.

Whether the erosion starts on forestland unwisely converted to
crops, or on arid grassland that is overgrazed and therefore fails
to reseed itself, the results will be equally tragic. When most of
the fertile topsoil is gone the land becomes unfit for any kind of
agricultural crop. The weathering of the ruined soil, the eventual
capture of the surface by pioneering plants, and the decay of plants
and animals will eventually restore the soil fertility—but only after
thousands of years. Perhaps some new civilization will make a
fresh start then and use the soil more wisely.

Fertility can be lost locally even where soil practices are good.
During great dust storms fertile tracts have been covered inches
deep with the sterile dust and sand blown from over-exploited
areas. Or carefully tended farms may be covered with soil wash
and pebbles from flooded areas upstream.

Sometimes "badlands" are created by industrial processes, as
from the poisonous fumes of copper smelters or steel plants. Strip
mining in Pennsylvania removes the fertile surface and then dumps
and buries it so deeply that the land becomes unfit for agriculture.
In Illinois and elsewhere, such stripped areas have been replanted
to trees, which can grow on land that will not support crops. The
trees slow erosion and stabilize the soil so that natural processes
can begin the slow renewal of surface fertility.

At the present time one-third of our total topsoil has been removed by erosion. Some of it went in steady week-by-week attrition. Much went in great floods. The 1936 flooding of the Ohio
River Valley removed 300 million tons of topsoil from the land—

about as much as the Mississippi River removes in a whole year. Some of the topsoil loss is widely distributed over the country, resulting in a shallower topsoil and lowered fertility. Some of it is concentrated loss, complete enough to rule out any agriculture. Of such completely ruined land there are 200 million acres—an area four times the size of Nebraska! We still have at present 450 million acres of good land. About one-fourth of this is receiving proper care by enlightened farmers; the remaining three-fourths is still being slowly or rapidly ruined.

Hindsight is much easier than foresight. And if our pioneer ancestors were blunderers who nearly ruined our heritage, at least they did so without access to the expert knowledge now available to us. If we do not act within the next generation or so to stop the steady washing of our irreplaceable topsoil into the rivers, there will be no sympathy for us from our unlucky successors.

We have already made a good start on research in farm practices, and some of what has been learned is already part of enlightened farm practice throughout the country. Contour plowing and strip planting are two of the most effective devices for delaying the runoff of water from the soil and so minimizing the load of silt that can be carried off. Terracing, which is practiced with great skill in Asia and in Europe, is being used in modified forms, on American slopes.

Quantitative knowledge of the effectiveness of different kinds of vegetative cover in controlling erosion can be obtained experimentally. Crops are planted in parallel strips on a hillside near an agricultural station. Water is then collected from below each strip, and the volume of runoff, together with the amount of contained silt, is recorded. Fallow soil is the worst from the point of view of erosion, corn and cotton a close second, grass and alfalfa fairly good holders of soil and water.

WATER SHORTAGES, DUST STORMS, and FLOODS

Common sense, which usually means the kind of sense that looks only at what is immediately at hand, tells us that water shortages and floods are two very different problems. Ecological sense, which looks beyond immediate effects and tries to trace out the interrelatedness of things says that water shortage and flooding are two aspects of the same problem—the effects of the widespread

This field is badly eroded and is rapidly becoming worse. It began to be eroded when the forest was cleared, the land cultivated for a few years, and then turned to pasture. As a result of overgrazing, the plant cover became so thin that its few roots were unable to hold the soil against the rain. Wisconsin.

A flood begins in fields, such as that above, where rainwater cannot be held in the soil, but rushes off so rapidly that the banks of the rivers cannot hold the added water. A flood, such as shown below in Idaho, is in large part a result of ecological misbehavior on the part of such people as the owner of the eroded pasture above. (Photos on this page and the next, by Ralph Buchsbaum.)

Overgrazing removes the leaves of the grass, prevents seed production, and leads to the loss of grass roots which hold the soil, not only against erosion by rainfall, but also against wind erosion. In addition, there are important biological factors which, if ignored, lead to destruction of grazing areas. In this New Mexico field there are too many sheep per acre for the grass to be able to maintain itself.

Cactus, invading a grassland, as in this field near Billings, Montana, is evidence of overgrazing. Removal of the grass makes for conditions favorable to the cactus. Recovery of the field is a long and costly process.

destruction of vegetative cover.

To the farmer who lives on the floodplain of a river, or to the engineer who is called in to save him from the consequences of living in a place whose very name suggests why he has no business being there, a flood is caused by too much rain upstream. To such short-sighted people flood control is therefore achieved by building up the banks of a river with levees so that the river cannot overflow its channel. This usually works for some years, during which water shortage may be acute, and then during really heavy rains the levees break and disaster overtakes everyone in the floodplain. Under somewhat better engineering theory the dry seasons and the wet can both be modified by building a dam with a reservoir that stores water in rainy seasons and releases water when needed during dry seasons. The difficulty with this is that it attacks the problem where it ends, not where it begins. If run-off of water into streams and rivers is not decreased by restoration of forests or improved farm practices, it increases from year to year as gullies work their way farther back into the valley landscape.

Erosion increases steadily, and a larger and larger load of silt is carried downstream to the reservoir. As the reservoir fills up with silt the speed of the river becomes checked farther upstream, the silt is dropped farther upstream, and sedimentation of the stream steadily moves upstream and into the tributaries. So the dam built to control flooding instead keeps extending flood damage upstream—in one case in Oklahoma as much as 147 miles upstream from the dam. The damage must keep increasing as long as nothing is done to hold the water where it falls—on the land itself.

Common sense tells us we must spend millions to build dams, but ecological sense tells us that their usefulness is in the best situations merely palliative, and in the poorly chosen spots actually a menace to the whole river valley. The ecological viewpoint calls our attention to the fact that in the very valleys that are flooded by "too much rain," the water table deep below the soil surface sinks lower every year. Large areas of the country are suffering from a drying up of wells. Municipalities must import water, by truck, from reservoirs at a distance. Overgrazed areas of the western plains suffer from floods one year and dust storms the next. The two problems are not caused by extremes of the weather. They are both caused by failure to maintain a land surface that can absorb rainfall and store it as water reserves below the surface, where it cannot run off to swell streams and where it will be available to plant roots whose steady growth holds the soil from blowing away.

To make matters worse, water use has increased enormously in the United States, and an expanding population and industrialization are making greater and greater demands for water at a time when our ecological sins are beginning to catch up with us on a big scale.

CROP PARASITES

If you have even a small backyard garden and have tried to compete with insect pests for flowers or for the tomatoes or beans you planted with great care, you will have some vague appreciation of how farmers feel about the four billion dollar crop loss they suffer every year from the insects that feed on plants. You will understand their desire to reduce this loss by an ever-increasing use of insecticides. The abundant food supplies on which we base our high dietary level in this country would have to be sharply cut back if we were to stop all use of poisonous chemicals on growing food. Food costs would rise steeply; our standard of living would fall. So it is very unlikely that we shall soon take so drastic a measure. Yet we are approaching a point at which the use of poisons is no longer just a mixed blessing, with limited risks, but a rapidly increasing trend which has multiplied the hazards to all animals, including ourselves.

With the introduction of the airplane as a device for spraying chemicals on a really gigantic scale not only over croplands but also over vast forest wildernesses that were formerly impractical to spray, the destruction of fish and wildlife, especially, has taken on menacing proportions. This aspect of the problem may soon receive some attention from the fishermen and hunters who together spend eight billion dollars a year on their sports and are becoming increasingly dissatisfied with the results. The hazard to human life has few spokesmen at present, especially since we are occupied with the even more threatening poisons of radioactivity. But it is not to be brushed aside lightly.

The city gardener who sees aphids on his rosebushes solves his problem in a direct and "common sense" way. He brings out a spraygun and removes a garden "enemy." To the gardener his problem is solved. But is it? The ecologist in the same situation asks himself what will now happen to the natural enemies of aphids, such as the lady-bird beetles. The answer is that they too are killed by the spray and that those who survive or escape the poison will find few aphids to feed on and will not multiply. Since aphid populations recover rather better than those of lady-birds because the former have special breeding habits which make this so, what will happen when the lady-bird beetles are not there to feed on aphids day in and day out, even during the weeks when the gardener is away on his vacation and fails to come around with his spraygun?

The farmer who sees insects feeding on his crops runs for his spraygun too. But there his "common sense" approach ends. And he often wonders why his problem doesn't end there too. The ecologically-minded observer could point out to him that plant pests, like all animals, are kept from overrunning the earth by many natural checks. Weather unfavorable to their breeding habits is one such check. Another is natural enemies like the birds. When insects are killed on a grand scale by poisons, the birds which feed on them must turn to other foods, often the grain of a farmer who has sprayed his vegetable crops. Now he drops his spraygun and reaches for a shotgun—"common sense" again. When he has killed enough birds the grain comes under even greater attack by the grain-eating insects that were formerly kept under control by the birds. At this point many farmers have left the farm for a less puzzling occupation in the city. But there are better answers to the farmer's dilemma. Some ecologically-minded farmers know them.

HUNTING AND FISHING

As we all know, fishing and hunting, while ever more popular, are becoming less and less rewarding both in catch and in the esthetic satisfactions of escaping temporarily from the city into the greenness of woods or the openness of a marsh or stream.

Fishermen long ago found that many of the major rivers and streams and even small creeks near big cities, had become polluted with sewage and with industrial wastes and no longer supported fish in large numbers. They solved this problem by travelling farther and farther from home to find unpolluted streams but soon these seemed after a time to become "fished out." Common sense said that the solution to this problem was to build fish hatcheries in which fish eggs could be hatched, and young fish raised. Then the fingerling fish could be put into streams and lakes that needed restocking. The results were not as expected. Fishing got worse and worse. Millions of fingerlings of the same size competed for food of exactly the same kind and size at the same time, and most of them did not live to grow up. The hatchery men had proceeded with no real knowledge of whether the lake or stream contained enough food to support so many fish, or whether the original adult population had really been "fished out" or had died of the same causes that were killing the young fish.

Many states are now abandoning fish hatcheries as being useless or even harmful. Instead, the states are asking their fish and game experts to learn more about what conditions make lakes and streams suitable for the breeding and growth of fish. One factor recently uncovered in a study of a "fished out" stream was that overgrazing by cattle had reduced the plant life on the banks of the stream and so removed important feeding grounds of the insects that formerly fell into the stream and served as an important part of the food of the fish. In some areas insecticides sprayed on crops were carried by heavy rains into nearby streams and lakes, poisoning the fish. So analysis of fishing problems requires an overall approach in which the nearby land areas must also be considered.

Friends of hunting have made many errors, and in some of them have been abetted by biologists who now know better. In Pennsylvania restrictions on deer hunting built up the deer population to a point at which none of the deer were well nourished, and in the first severe winter large numbers perished of starvation. The same end result has been achieved many times in the West by unlimited killing of mountain lions and coyotes that formerly kept the deer in check.

We shall have to learn to ask whole batteries of questions. What do deer eat when they have plenty of food? What do they add to their diet when food runs low? How many deer can be adequately fed in a forest of a particular size? At what size of population will they begin to damage the forest by excessive browsing? Do their breeding habits suggest that both sexes should have the same degree of partial protection? The questions are endless. But so are the unhappy problems of people who stop asking questions.

Chapter 2

THE PHYSICAL ENVIRONMENT

DESPITE recent successes in probing outer space, our green little planet is still home to the only kinds of life we know about. That part of earth which supports life we call the biosphere. Hardly more than a surface envelope, it includes the soil to a depth of relatively few feet, all the oceans and fresh waters, and the atmosphere. Wherever in the biosphere we find plants and animals living, their presence attests to the fitness of their surroundings, or environment, to support the kind of life found there.

Often the presence of some plant or animal is a more sensitive indicator of a particular set of conditions than even the most careful measurements, made with the best scientific instruments we now have. The biologists at the Marine Laboratory in Plymouth, England, have been trying for a long time to work out a reliable method for predicting good and lean fishing years and for directing fishermen to the best areas in each year. So far the most convenient indicator has proved to be an almost invisible, transparent arrow-worm, Sagitta, which abounds in surface waters and is an important part of the food of small fishes. Sagitta setosa is found in waters low in phosphates and other nutrient salts, and Sagitta elegans is abundant in phosphate-rich waters. We know that phosphates promote the growth of the microscopic green plant life, the plant plankton, that we often call the "grass of the sea" because it supports all the other life of the oceans. The green plants are fed on by countless billions of minute crustaceans. And these in turn are devoured by sagittas and by small fishes, both of which then fall prey to herrings and other larger fishes. All the while the fishermen wait impatiently, nets ready on deck, for the schools of herring to appear. On their behalf the biologists, sampling the waters, try to relate good fishing areas to physical and biological factors. Working out of small boats, they find it much more convenient to examine the inch-long sagittas and determine their species, than to analyze the phosphate content of the water or to make time-consuming counts of the number of microscopic plants.

All of this suggests that the roles, and also the exact numbers of plants and animals, are inseparably bound up with the numbers of other living things, as well as with the amounts of materials and with the kinds and magnitudes of the physical forces acting at any time. The plants cannot grow without phosphates, but the phosphates would soon be all locked up in the bodies of dead plants and animals if they were not decomposed by bacteria on the ocean bottom, and the phosphates returned to the water to be reused. These ceaseless exchanges of materials and of energy between living things and their environment follow circular pathways, which are repeated in endlessly repetitive cycles. Such cycling systems are called ecological systems or ecosystems, and they can occur at any level of biological complexity. The ecosystem that keeps a single individual alive and self-renewing is the chief interest of the physiologist. Most often the ecologist concerns himself with large, self-contained units, like a desert, a prairie, a forest, or a lake, that require little replenishment from the outside, except for sun and rain. Of course, the most completely self-contained ecosystem is the biosphere, but its size and complexity require that it be studied a part at a time.

For classroom study a simple and convenient ecosystem would be a "balanced" aquarium with a cover. If it contained all of the basic components of an ecosystem (page 62) it could continue to cycle materials and energy for a very long time without renewal from the outside. In nature or in a balanced aquarium, the non-living and the living factors are all interdependent and interacting, and not always easily separated into the two categories given below. A snail may move about on the surface of the glass or on a living plant. A plant may receive its carbon dioxide from one of the fishes, from another living plant, or from the decomposition of a dead plant or animal. A fish may die because of a sharp rise in the temperature of the water, or because the rise in temperature killed another living thing upon which the fish was feeding. Nevertheless, it is difficult to clarify the relationships of living organisms to their environment without lifting them out of their complex natural setting and discussing them one at a time. So we shall temporarily divide the factors of the environment into two kinds: (1) the non-living factors, or physical environment, including the medium and substratum, heat, light, water, chemicals, gravity, etc. ; and (2) the living or biotic environment, which consists of the plants and animals themselves (Chapter 3). Later in our story we shall tie them together again.

Only a few of the physical factors will be discussed and illustrated with respect to their action upon living organisms. And these will be considered from only two main viewpoints: how plants and animals are adapted to fit their environments, and how organisms are limited in their distribution.

To understand what is meant by a "limiting factor," let us take this simple example by analogy. If, in the manufacture of automobiles, a certain part is essential, then regardless of the superabundance of all other parts and the working capacity of the factory, automobiles will be produced only at the rate at which that particular part can be produced. Such a part constitutes the limiting factor in the production of automobiles. Similarly, no matter how much carbon dioxide, water, oxygen, and mineral salts are present in the deeper parts of the ocean, green plants cannot grow there because of the absence of light, the one necessary additional factor for green plants on the ocean bottom. Usually, there are many factors in any given case. Deserts are not only too hot but also too dry for most plants and animals, and the soil too alkaline. Organisms dependent upon a set of certain factors or combinations of factors are limited to those areas where they occur in adequate quantities; we refer to this as the principle of minimum conditions. But living things also are barred from certain areas by too much moisture, too much light, too much heat, or perhaps by too much of certain chemicals (e. g. , selenium) in forage plants. Nor can they persist where there are too many competitors or enemies. Hedged in by maximum limits as well as minimum ones, they are limited in their distribution and in the success they attain by their limits of tolerance for various environmental factors.

Of the many physical factors affecting living organisms, only a few will be considered. Some factors are more important than others; and organisms vary with respect to how much they are controlled or limited by the various physical factors.

MEDIUM and SUBSTRATUM

Though plants or animals may live in mud, sand, soil, salt water, fresh water, rock, wood, blood, living tissue, and in many other habitats, including even oil, their bodies are always completely surrounded by a film of gas or liquid—usually air or water with which they maintain respiratory exchanges. A watery or aquatic habitat usually contains dissolved gases of the air. The terrestrial habitat always has some water in its atmosphere of air.

A substratum is a solid surface to which any organism attaches itself, or on which it rests or moves freely. And it is any solid material in which an organism lives or into which it extends part of its body. Aquatic animals, even very large and heavy ones, may float or swim, using the water medium for mechanical support; or they may live their whole lives resting on or burrowing in the bottom of the ocean or lake. Air is not so buoyant as water, and few terrestrial animals can live suspended in air for long periods except microscopic bacteria, molds, and spores. Birds, bats, and insects are adapted to use the air effectively for mechanical support, but even they must spend most of their lives at rest on some solid substratum.

Plants may have structural adaptations to the particular substratum on which they live. Most familiar to everyone are the root systems of almost all ferns and seed plants which enable them to maintain themselves in a favorable position in their environment. Such plants as the morning glory, the grape, and the English ivy have special modifications of the stem which enable them to climb up supports and expose more leaf surface to the light. "Spanish moss" is adapted to live hanging from supports above the ground and is usually found on live oaks or even on telephone wires in humid areas in the South.

Many of the large marine algae have bladder-like enlargements of the stalks that keep the blades afloat near the surface of the water, where light is most intense. In most the stalks are firmly attached to the rocky substratum; but the alga called "sargassum weed" lives as large free-floating clumps over an extensive area in the Atlantic Ocean off the southeastern shores of the United States. Almost any floating clump supports tiny flatworms, snails, worms, shrimps and crabs that are themselves unable to float. Most interesting is the sargassum fish, which has a special pair of holdfasts on the head and modified front fins for holding on to the alga. The fish can release itself, swim about to feed, and then return to its floating perch. Seed plants such as the water hyacinth are able to grow on the surfaces of ponds or streams, floated by air-filled leaf stalks. We are all familiar with the "duck weeds" and water lilies that float on the surfaces of ponds by means of flattened leaves.

Prop roots of mangrove help to support the plant in soft soil in water. The roots also afford a substratum upon which marine invertebrates can attach (as on the wharf piling on the opposite page). Florida Keys. (Photo by Ralph Buchsbaum)

Penguin, showing adaptations to water: streamline form, wings adapted for swimming; webbed feet. Bristol Zoo, England. (Photo by Monte Buchsbaum.)

Dolphin, as seen from deck of a ship. These are mammals adapted to aquatic life. Note the streamline form. Gulf of Mexico. (Photo by Jan Hahn, Woods Hole Oceanographic Institute.)

Streamlining for burrowing in the soil is shown by this legless amphibian (a cecilian) from the tropical forest of Panama. Blind and wormlike, it feeds upon worms and insects. Cecilians breed in the soil. 14 in. (Photo by R. Buchsbaum)

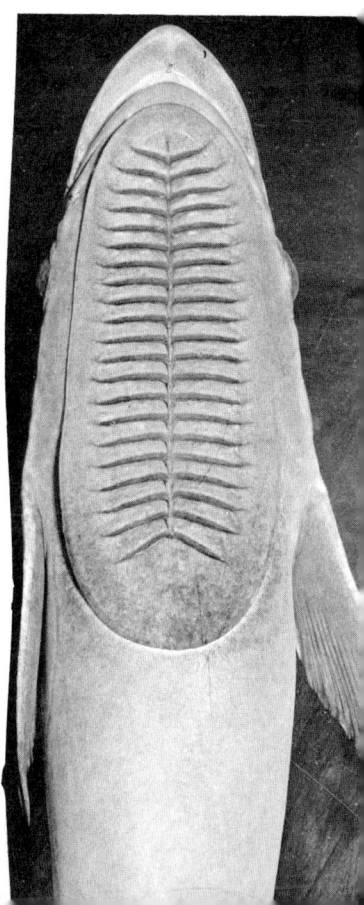

Substratum may be a limiting factor for sessile marine organisms such as barnacles and mussels. They are able to grow in this sandy beach only on the wharf piling. LaJolla, Calif. (Photo by R.B.)

Shark sucker is a fish adapted to "hitch-hike" on larger animals such as sharks or turtles by means of a sucker on top of its head. It is shown here in an aquarium attached to the glass. Key West, Fla. (Photo by Ralph Buchsbaum.)

Animals may have structural adaptations to the particular substratum on which they live. The streamline form of fish, seals, and whales, the streamline form of fleas and also the combs of the flea which prevent slipping while moving through hair, the flattened bodies of lice, the long legs of wading birds, the sharp claws that enable squirrels to run up the trunks of trees, the well-protected hoofs of horses, zebras, pigs, and camels, the exceptionally large paws that help to keep the Canada lynx from sinking too deeply in the snow, the hooks that enable the tapeworm to cling to the intestinal wall instead of being carried out with the moving stream of intestinal contents—all these are successful adaptations to a variety of types of substratum.

While some organisms have become structurally adapted to their substratum, others have changed the substratum to fit themselves. Examples are the nests of ants, the elaborate tunnels and galleries of termite nests, the tubes of many marine annelids, and most obvious of all, the streets, bridges, tunnels, airplanes, buildings and homes of man.

Substratum may be a limiting factor. An organism that has become well adapted to a specific type of substratum is limited in its distribution by the availability of that substratum. Off the northern California coast there are regions where large brown algae grow crowded together, while a short distance away there are none at all. The blades of the alga float, byoyed up by air bladders, and the alga is kept from being washed up on the shore by a long stalk and a holdfast attached to a rock at the bottom. If all other conditions are alike in the two regions we can be fairly sure that the substratum of the region where the algae are not found is not rocky but probably muddy or sandy, and the algae cannot attach themselves.

The sessile (fixed) animals of the rocky ocean shore, such as the barnacles, anemones, and tube-worms compete not so much for food as for a piece of substratum on which to settle down. Any animal that succeeds in establishing itself on a suitable spot on a rock can always obtain enough food by straining organisms from the water. The sea water near shores is a living "soup," and there is plenty for all, but one needs a solid vantage point from which to do business. While substraum is likely to be an important limiting factor for sessile animals of rocky sea shores, it is not the only one. Many submerged rocks on the sea shore are quite barren because they are exposed to wave action that is too strong for animals to survive there. Or, for certain animals there is too little wave action to provide enough food or oxygen. Or, there is too much exposure to air at low tides. Or, for animals adapted to conditions of the upper shore, there may be no crevices in which to hide. So that while a rocky substratum may be an indispensable condition for the attachment of a barnacle, no matter how favorable the other factors, these other factors all operate in limiting a particular species of barnacle to a particular part of a rocky substratum.

This reminds us again that organisms live in an environment where many factors are acting. It is most frequently true that animals and plants are limited not by a single factor but by a whole complex of factors. However, the ecologist is always seeking out, for the purposes of study, those cases where only a single factor is predominant.

HEAT

Organisms differ in their ability to tolerate ranges of temperature. In general, the lower the organism in the scale of evolution, the better is its ability to withstand extremes of temperature. Some bacteria can survive exposure to -454°F. for several months. This is the temperature of liquid hydrogen and also, perhaps, of interstellar space. The survival of bacteria at such low temperatures supports the contention that organisms may be brought here on meteors—though not commonly, for such bodies usually reach too high a temperature for the survival of life. Colpoda (a ciliated protozoan) was kept for one week at -315°F., and when warmed up grew as well as control animals which had been kept at room temperature during all of this time. Land snails have been cooled to -184°F., and the Promethea moth pupa to -31°F. without injury. Frogs have been taken down to -18°F., and fishes to -4°F. If such frozen fishes are dropped on the floor they break in two like a cake of ice; yet if handled carefully and warmed slowly they are none the worse for this treatment. The survival of animals at temperatures below the freezing point of water is related to the phenomenon of "supercooling." Liquids may be cooled below their freezing points without solidifying; but if a crystal is introduced or the supercooled liquid is jarred, crystallization will set in with great rapidity. An insect, cooled without crystallizing, will recover if thawed out properly. Any rough treatment results in immediate crystallization and death.

Many simple forms can tolerate relatively high temperatures.
Some bacteria can be heated to 284°F. momentarily and survive.
But none can withstand 320°F. for one hour; hence this is used as
a standard for dry-sterilizing surgical and bacteriological apparatus. Steam sterilization is accomplished in 15 minutes at 15 pounds
steam pressure; the temperature that develops is 257°F. Certain
blue green algae live in hot springs and pools at 190°F. Beetle
larvas living under bark at 126°F. is a record performance for
insects. However, about 118°F. is the highest temperature that
can be tolerated by most forms. Death at high temperatures is
thought to be due to changes in the proteins or fats.

So far we have considered only "cold-blooded" animals. "Cold-
blooded" is really a poor term because during the summer the
blood of a cockroach, for example, may be warmer than that of
the woman whose kitchen it frequents. It would be better to speak
of the cold-blooded forms as animals which do not have a temperature-controlling mechanism, or poikilothermal animals; their
body temperatures, consequently, vary with the external conditions.
In temperate regions most poikilothermal forms are dormant all
winter. In the spring their temperatures rise with the external
temperature and they resume activity. A few of the invertebrates
have special, though not very effective, devices for maintaining a
temperature higher than that of their environment. In winter,
honeybees arrange themselves in a close cluster and those inside
constantly vibrate their heavy wing muscles, thereby liberating
enough heat to raise the temperature of the hive several degrees
above that of the air outside. In summer the bees can lower the
hive temperature by creating a current of air which increases
evaporation.

Plants do not have special temperature-regulating mechanisms,
but in cool weather the body of a plant may have a higher temperature than its surroundings as a result of heat liberated by respiration. In summer the evaporation of water from the leaves exerts
a marked cooling effect.

Most poikilothermal animals cease to be active when the external temperature goes down to about 43°F. or rises to about 108°F.
But within these limits, higher temperatures speed up metabolic
processes and increase activity. Ants run faster in summer than
in spring. Herring eggs develop in seven weeks at 33°F. and in
one week at 61°F. Crickets chirp at a lower rate in cool weather

and at a higher rate in hot weather. One investigator found that he could estimate the air temperature by counting the rate of chirping: 47 chirps per minute indicated a temperature of 49°F., 71 chirps per minute meant 55°F., and 150 chirps indicated 75°F.

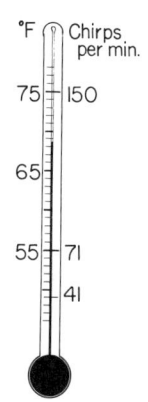

Cricket thermometer.

There are temperatures too low and temperatures too high for any living activity. There is also a temperature range, for any given species, within which it can maintain life but with limited efficiency. The temperature at which an organism best maintains itself may be termed its <u>optimum temperature</u>. The optimum temperature varies with other physical factors. Within limits, the optimum temperature for photosynthesis varies directly as the amount of light and the supply of raw materials; that is, at 90°F., much more CO_2 and water are required for the optimum rate of photosynthesis than at 60°F. For man, the optimum temperature depends upon other factors, chiefly the water content of the atmosphere, that is, the humidity. One feels fairly comfortable in a room at 70°F. provided that the air is about 50% saturated. If the room is exceedingly dry, so much evaporation of the skin occurs that one feels cold. Similarly, when the humidity is extremely high, one may feel uncomfortably warm at 70°F. Tuberculosis bacilli grow better at about 100.4°F. than at lower or higher temperatures. In general, the conditions which result, not in the fastest or most intense biological reactions, but in the greatest chance of survival of the individual or of the species, are the "optimum conditions."

Animals which maintain themselves at their optimum temperature practically all of the time are the "warm-blooded" or <u>homo-thermal</u> animals, the birds and mammals. Their life-processes are so delicately adjusted to function at their normal temperature that if their temperature-controlling mechanism fails to function properly, death results. In prolonged exposure to cold the body temperature of man may fall from 98.6°F. to somewhere near 65°F., and at about that point (lower or higher, depending upon other circumstances) the heart fails; in sunstroke, the body tem-

perature at the time of death may be 112°F. However, while birds
and mammals cannot withstand very serious changes in their internal temperatures, they are relatively independent of changes in
the external environment; they can maintain a high level of activity
at all seasons of the year. In temperate and in Arctic regions the
poikilothermal animals are practically all dormant during the winter; the only active animals are birds and mammals.

During the winter some mammals, too, become dormant, that
is, hibernate. At this time the body temperature (normally about

Bear awakened from deep winter sleep, a dormant state less profound
than hibernation. The opening to its chamber was partly concealed by
a log which was moved aside for this picture. (Photo, Penn. Game Com.)

100°F.) falls to 45°F. or even lower. The heart beat of a ground
squirrel drops from a summer average of over 200 beats per minute to a hibernating average of less than 20 beats per minute. The
respiratory rate of the dormouse falls from a normal rate of 80 to
a hibernating rate of 10 inspirations per minute.

Many marine animals also maintain a constant level of activity
throughout the year. Because of the high specific heat of water and
because of the tremendous volume, temperature fluctuations in the

oceans are not as great as on land. In many places, deep ocean animals have a nearly constant temperature of about 39°F. throughout the year and have become so adjusted to this temperature that they do not live very long if put into warmer water. And, even at this relatively low temperature their physiological processes are carried on at a rate not too different from their relatives in warmer water. This is accomplished in the cold-water forms by the evolution of enzymes which enable the animals to metabolize at a high rate.

Temperature may act as a limiting factor. The polar bear is adapted for living in very cold climates by its heavy insulating coat of fur and fat. The distribution of the polar bear is limited to regions in which the temperature averages no higher than 32°F. Other factors, such as the availability of food, are important in the life of the polar bear and further restrict its range. On the other hand, man and his primate relatives are adapted for living in very warm climates by their thin coats of hair, relatively little fat, and their efficient cooling mechanisms. Among all the primates, man alone is not limited in his distribution by temperature— but this is not because man is adapted to living in cold climates, but because he has produced a nearly tropical climate in his home.

The malarial parasite in its alternate host, the Anopheles mosquito, does not develop when the temperature falls below 77°F. Hence, the distribution of malaria is limited to regions which fulfill the temperature requirement. But a man who acquires the malarial parasite from the bite of a mosquito in a warm climate can then go to colder climates and continue to harbor the parasite in his thermostatically-controlled warm blood until the end of his life.

Some plants are limited by the average yearly temperature, while others are limited by the extremes of temperature, because they cannot survive freezing weather even for a brief period. Certain plants can grow in regions where the temperature is too low for the ripening of their fruits. The banana plant is used extensively in Los Angeles as an ornamental garden plant, but it does not fruit because of the low temperature. Palms in Los Angeles produce large numbers of immature fruits which never ripen, but prosperous date palm farms exist in the warmer valleys not very far away.

Temperature may have a marked effect on the appearance of organisms. The Himalayan rabbit is mostly white but has black ears, tail, and feet. If the hair is plucked from a part of the white area on the back, it will grow in white again if the animal is kept at room temperature (i. e., 68^OF.). The new hair grows in black if the rabbit is kept below 50^OF. Conversely, if hair is plucked from a black region and the plucked patch is kept warm with a bandage, the new hair will grow in white. The color pattern of

Himalayan rabbit. (Photo by Monte Buchsbaum)

certain animals is thought by some to have adaptive value, as it may render the animal less conspicuous. Thus the ermine is white in winter and light brown in summer. It is thought by others that this change in color is not necessarily adaptive, but (like the Himalayan rabbit in the experiment) is merely a reflection of metabolic changes related to the warmer weather or other factors. The solution of this problem awaits very careful experimentation.

LIGHT

Light may be a limiting factor. The penetration of light into the ocean determines the depth at which green plants may grow. Since CO_2, water, and salts are present in abundance, light is the most important limiting factor. Algae are most abundant at the surface of the sea and decrease with increasing depth until at 200 feet the plant life is very sparse. And since less than 0.1% of the light reaches below 600 feet, green plants are absent from the ocean depths. Of course, the absence of plants means that animals that depend directly upon plants for their food supply are also absent. Only carnivores and scavengers live on the ocean bottom, and these feed on each other and on organic stuff which settles out from above.

In dense forests the tall trees spread their huge crowns of leaves and take most of the light, shading the ground below. The floor of the forest has a soil rich in salts and well-permeated with water, but because of the little light that it receives it supports only plants which are adapted to photosynthesize at relatively low light intensities. Algae, mosses, ferns, and some seed plants manage to grow under these conditions. Thus extensive areas of the earth's surface support less than their possible quota of life chiefly because of the lack of the single factor: light energy.

Light affects animals indirectly through the fact that animals depend upon plants for their food. Light also affects animals in a variety of other mechanisms. Exposure to ultra-violet radiation enables some animals to manufacture vitamin D. The shortening of the duration of daylight in the fall is the chief stimulus to begin the migration south of certain birds. Some of these birds can be experimentally inhibited in this behavior if they are exposed to an artificially lengthened day.

Light affects animals directly as they perceive their prey or detect their enemies through the sense of sight. Light-sensitive organs are present in nearly all animal groups from the protozoans to the highest forms. Many animals have evolved structures or forms which enable them to escape detection by their enemies which hunt by means of sight. Examples are the walking-stick insect, the dead-leaf butterfly, and the various color-patterns on insects, birds, snakes, and mammals. Mimicry and concealing coloration can be thought of as a counter-adaptation to meet the adaptation of sight developed by carnivorous animals. The brilliant coloration of certain birds is likewise dependent upon sight and is an adaptation largely based upon sexual selection.

In caves there is total darkness. Among plants only bacteria and fungi can subsist and these do so on organic material that enters the cave from the outside. Animals that take advantage of the protection offered by a dark cave must either leave the cave to feed or must make the best of the organic material that is brought in. Of the animals that go out to eat, the bats are the most interesting. A large cave may harbor thousands or, like the Carlsbad Caverns in New Mexico, even millions of bats, which usually leave the cave after sunset, and, after spending the night flying about feeding upon insects, return just before sunrise. Their ability to fly about in total darkness without striking objects is a remarkable adaptation to a lack of light. Bats avoid objects by constantly emitting high-pitched shrieks and reacting to the echos. Also, the wings are sensitive to slight differences in air currents. Among the animals that never leave the cave are crickets, beetles, crayfish, flatworms, and fish. These feed on animals that stray into the cave, are blown in, or are washed in by water that runs into the cave. Most cave animals are white and blind, presumably because defects in vision and in pigmentation are not a handicap to a cave animal, as they would be to an animal living in a lighted environment, where its competitors and enemies could see. When animals with defects are not at a disadvantage, they are not eliminated by natural selection. And over many thousands of years the defects accumulate and are spread throughout the population until such capacities as vision and pigmentation may be lost completely.

WATER

The adaptations of plants and animals to life in the water are too numerous to be discussed in this brief account. Instead, attention will be directed to water as a limiting factor—how organisms are adapted to a scarcity of water.

Most aquatic animals cannot live without a large supply of constantly renewed water. Such are most fish and other truly aquatic forms, whose respiration is adjusted for utilizing oxygen dissolved in water. These must live in large bodies of water or in swift streams. But certain aquatic types like lungfish, salamanders, and crayfish survive temporary drying by burrowing into the damp mud. Frogs spend much of their adult lives on land but cannot live very far from the water, to which they must return to lay their eggs. Also, frogs have a thin skin which must be kept moist, and so they must live in regions of relatively high humidity. Garden toads have solved this difficulty to some extent by the development of a leathery, waterproof skin which greatly reduces loss of water by evaporation and enables them to invade regions not open to the

Cave salamander is white, as are practically all cave animals. This salamander has some tiny pigment spots and eyes, but many such animals are blind. From a Missouri cave. (All photos by R. Buchsbaum)

Cave fish, from an Indiana cave, are white and blind. Defects in pigmentation and in vision are not a handicap in the darkness of caves and so are not selected against. Length of fish, 2 inches.

Cave flatworm, *Sorocelis americana,* from a cave in the Ozark region. It is ½ in. long, white, and has numerous eyes (which are said to darken if the animals are kept in the light).

Crayfish, white and blind, from a cave in Missouri. The body was 2¼ in. long; the antennas were 3¼ in. long. Blind cave animals frequently have good tactile sense organs.

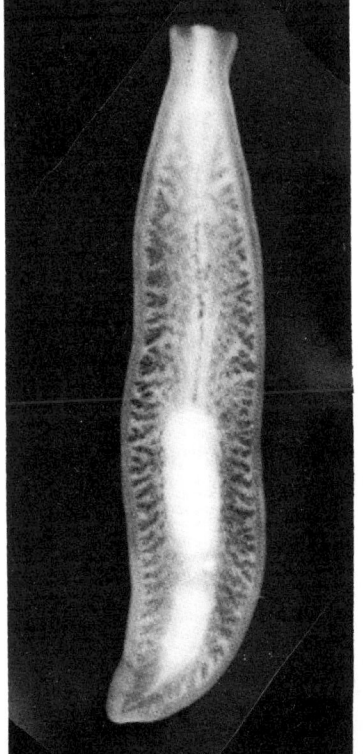

frog. But reptiles are the first vertebrates to have evolved toward
eliminating water as a limiting factor in their distribution. Reptiles lay land eggs and have a heavy scaly skin through which
water cannot escape. Some reptiles (turtles and crocodiles) have
secondarily returned to the water, but the reptile group as a whole
achieved its early success through adaptations to lack of water,
and the group now has a large number of the most successfully
adapted desert animals.

In the arid Southwest, where animal life is scarce, snakes and
lizards are relatively abundant. The horned lizard is an excellent
example of an animal which has become completely adapted to life
under the most rigorous conditions of drought. The heavy skin
prevents water loss; the excretions are all dry; and the supply of
water comes from minute amounts of water contained in the tissues of its prey plus the water resulting from metabolism.

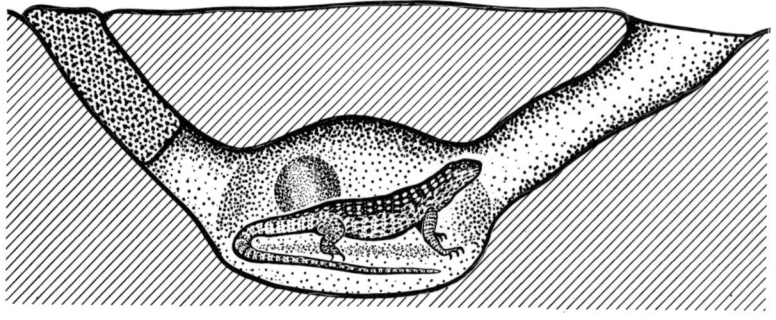

Lizard, Dipsosaurus, in its burrow. Surface sand temperature was 107.5°F.;
temperature at the bottom of the burrow was 95°F. Near Palm Springs, Calif.
(After Norris.)

Other examples of animals which can live on their metabolic
water are the kangaroo rat, the clothes-moth, and the meal worm
(beetle larvas frequently found in cereals). Many of the large mammals of the arid regions of Africa can travel hundreds of miles
from their nearest source of water; an occasional big drink lasts
them a long time. Thus we see that animals, which can move
about readily, are not necessarily seriously limited in their distribution by a lack of water to drink. Since animals are dependent
upon plants for food, however, most plant-eaters are restricted
in their range by water acting as a limiting factor for their food.

Kangaroo rat from the desert in southern California. Its body is 5 in. and its tail 9 in. long. Herbivorous and nocturnal, it hops about actively on its long hind legs. (Photo by Ralph Buchsbaum.)

Since plants cannot move about to find water and are closely limited by the moisture in the soil, the adaptations of plants to arid conditions are necessarily more thoroughgoing than are those of animals. Ferns have developed very little resistance to drying and are restricted, the world over, to moist regions. In the red-wood forests of California, where ferns grow abundantly, there is practically no rain in summer, but heavy fogs serve as an effective substitute. Mosses are more successful than ferns in solving the problem of water scarcity. Seed plants are still more adaptable. In addition to their independence from external water for transporting sperms in fertilization, they have made other striking adjustments to the lack of water. These are of two general types. They may increase absorption by the development of extremely long roots which penetrate to deeply situated or widely scattered water-containing layers. Some desert plants have roots that pene-

Plants adapt to decreased water supply by shedding some of their transpiring organs. (Parts shed are shown in black.) 1. Whole shoot shedders, about 85 % of the total flora, discard everything above ground. 2. Leaves shed in mid-summer. 3. Large winter leaves discarded, the plant develops small summer leaves. 4. Jointed-branch shedders discard part of last year's branches. See also the desert section in Chap. 8. (After Zohary.)

trate as deeply as forty feet. Or they may have aerial parts which
are modified to reduce evaporation. The second method is much
more common than the first and is effected in many different ways.
The breathing pores (through which water is usually lost) may be
reduced in number, or they may be protected by hairs from the
drying action of the air, or they may be sunk into grooves or pits
(as in cacti). The leaves may be rolled up (as in grasses) or they
may be reduced to mere spines (as in cacti). In these cases the
stem usually assumes the photosynthetic function. In addition,
most cacti have a thick leathery epidermis and special water-
storing tissues. Like the desert animals, plants which live on a
low-water economy have a relatively low metabolic rate, in spite
of the high temperature. Hot-house cacti which are watered relatively frequently grow very much faster than do those in nature.
Under these artificial conditions several cacti can be grown in one
pot. In the desert, however, although sunlight, CO_2, oxygen, soil
minerals, and all other factors are present in abundance, water
acts as a limiting factor in excluding most plants and in determining
the extremely wide spacing of those plants able to survive the arid
conditions.

Desert plants, like the Ocotillo in the foreground, have many adaptations for
conserving water, but are spaced as close together as the water supply permits. The Ocotillo here in April is in bloom before the tiny leaves appear;
southern California. (Photo by R. Buchsbaum.)

Chapter 3

THE LIVING ENVIRONMENT

ALL LIVING plants and animals affect each other directly or indirectly, and collectively constitute the living or biotic environment. In this chapter we shall discuss only a few of the main ways in which living organisms of different species interact with one another: plants with plants, plants with animals, and animals with animals.

Among biotic relationships, the one that is of primary importance is that of the struggle for energy. It has been said over and over again that plants make their own food, but this does not mean that they can make their own energy. They must capture their energy from the sun. Among plants, the struggle for energy is chiefly a competition for a place in the sun. And plants differ greatly in their adaptations for using available light.

Trees growing alone in a field are likely to be wide-spreading, symmetrical, and the lowermost branches are alive, frequently close to the ground. Very different are those of the same kind growing close together in a dense forest; such crowded trees are likely to be tall, the branches may be short, and the lower branches frequently are dead, because not enough light reaches their leaves to keep them alive and growing. Ferns and other small plants are adapted to conduct photosynthesis at a very much lower light intensity. If all other conditions for plant growth are abundantly supplied, the number of plants is limited only by the surface area which can be exposed to light. Looking down upon a bed of clover or on a patch of ivy, or on a tropical jungle (from an airplane) one sees a solid field of green growth, in which plants and their leaves are arranged to secure the most advantageous light exposure. If one species of plant tends to cut off the light from others by more rapid growth over the ground or by growing taller and holding its canopy of leaves over the others, this species becomes the dominant species. The beech or maple tree in a woods, rushes in a marsh, grass on a prairie are examples of plant dominants.

Elm tree, alone in open field, grows symmetrically. Ohio. (Photos by R. Buchsbaum.)

Two elm trees, growing close together, have relatively few branches on their facing sides. Ohio.

Many of the tree-climbing vines of temperate and tropical forests are adapted to very low light intensities and thrive only in the shade of forest trees. The various species of Philodendron, which are so popular as house plants, are climbing vines from tropical forests, and that is why so many species thrive on living-room mantels, but turn yellow when their well-intentioned owners put them in sunlit windows. However, the climbing of the ivy or morning glory is an adaptation for better exposing leaves to the sunlight. All these climbing plants are rooted in the ground. In the tropics, where the air is very humid and water is thus available from the air, some plants, among them many orchids, are able to live perched high on the branches of trees where they receive more light than they would on the ground; such plants are known as <u>epiphytes</u>. Another epiphyte is the Spanish "moss" a highly specialized seed plant which grows hanging in long festoons from trees and other supports in our own South and West. The epiphyte manufactures its own food and does its host no noticeable harm, so the relationship is not one of parasitism, but rather of the kind we call <u>commensalism</u>. A commensal gains some benefit from its host without conferring either harm or benefit.

Mistletoe growing on tree in France. (Photos by R. Buchsbaum)

Spanish moss on cypress; Florida.

Epiphyte, growing on tree trunk in a dense tropical forest. Panama. (Photos by R. Buchsbaum.)

Indian Pipe, a white, saprophytic seed plant. Beech and maple forest. Provincetown, Mass. Plants, 5 in. high.

Not all plants are able to make their own food if supplied with light. The mistletoe grows perched high on the branches of various trees, not merely as an epiphyte but as a <u>partial parasite</u>, for though it has green leaves and can carry on photosynthesis, it obtains water and soil salts from its tree host.

True plant <u>parasites</u> derive all of their nourishment from their plant hosts, and so are not dependent at all on a favorable exposure to light. Such are the fungi like wheat rust and certain mushrooms.

Plants which live on organic material from dead plants or dead animals can live in deep shade or even in dark caves. They are referred to as <u>saprophytes</u>. Examples are many mushrooms and the colorless seed plant called Indian Pipe.

Some plants live together, not in competition, but in a relationship which is mutually beneficial, and which is known as <u>mutualism</u>. Good examples are the lichens, which consist of a fungus mycelium enclosing in its meshes, algal cells. The fungus derives its food from the algae. The algae are protected within the meshes of the fungus mycelium from excessive drying. Both organisms are thus able to live in places, such as bare rock, which cannot be inhabited either by fungi alone or by algae alone, or by any other organisms. (Mutualism is sometimes erroneously called "symbiosis." The word <u>symbiosis</u> means "living together" and is best used as an inclusive term to cover commensalism and parasitism, as well as mutualism.)

Plants and animals interact in a variety of ways, and even the

least ecologically-minded of us knows that animals eat plants.
Tracing out the exchanges of energy-bearing substances is one of
the major tasks of the ecologist. Oddly enough, it is not always
easy, even for relatively common animals. Ecologists cannot go
around stalking every kind of animal to see what it eats during all
hours of the day or the night. They have had to depend in part on
people who live close to wild animals, like farmers, and the farmer is often a very biased witness. To hear many farmers tell it,
coyotes live exclusively on sheep, foxes on chickens. When ecologists have had time to look carefully into the food habits of these
animals they have found that coyotes live largely on small rodents
while foxes have a varied diet of rodents, berries, and other
things. Many animals are adapted, as we shall see, to a more or
less specialized diet, but animals may be hungry when their favorite food is not ripe or not present. Animals, frequently of necessity,
eat whatever is available.

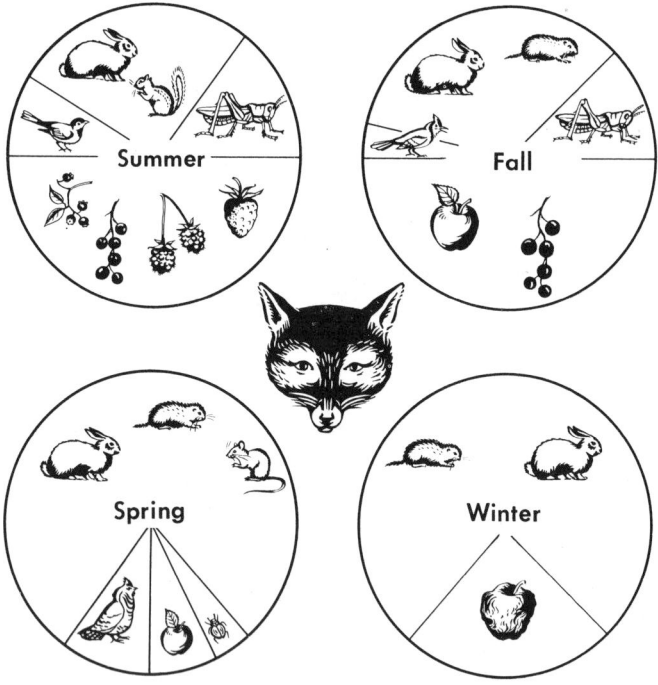

The red fox has a varied diet. What an animal eats depends partly upon its special
adaptations, but also upon what is available. As can be seen from this diagram, the
fox is not the pure carnivore expected on the basis of its sharp teeth and its predatory reputation. (Based on Cook and Hamilton.)

Even though animals may, at times, depart from eating a particular food, or type of food, animals are generally structurally adapted in their feeding mechanisms. Animals that eat plants are known as <u>herbivores</u>. They may be smaller than insects or as large as elephants. Among microscopic herbivores are some rotifers which feed upon unicellular algae and some of the copepods in the sea which feed largely upon diatoms. Many insects are adapted to sucking juices from plant leaves and stems. Other insects have mouth parts adapted for chewing leaves. The leaf-cutting ant goes a step farther. It takes the cut-off pieces of leaves into an underground nest and uses them as a medium on which to grow fungi which form its basic food. Snails have a rasping organ, the radula, which scrapes off small shreds of plant material upon which the animals feed. Ungulates have highly modified grinding molars adapted to breaking down grasses and other plant parts. Some of these animals have elaborate modifications of the stomach so that the cellulose of which plant tissues are composed may be digested.

Many of the relationships between plants and animals are adaptive for one member but merely accidental for the other and of no special benefit or harm. A man walks through a field and picks up burrs which drop off later in another place—thereby providing for seed dispersal. Some seeds can pass through the digestive tract of an animal and still germinate in whatever spot they have been eliminated by the animal. Others are so adapted that they cannot germinate unless they have first passed through the digestive tract of an animal. One such tough-coated seed in Africa must be scarred by the teeth and pass through the gut of a wild boar before it will germinate.

Aphids on the stem of a rose bud. The aphid inserts it stylets into the vascular bundles and sucks out plant juices. Winged forms arise in the fall and provide for distribution to new plants. (Pittsburgh, Pa. Photo by R. Buchsbaum.)

Plants modify the climate and provide particular physical conditions for animals. For example, tree frogs which have moist skins can live only in regions of high humidity such as is provided in a moist forest. We plant trees (or, would be wise to do so) to shelter our homes and streets from the full impact of the summer sun. It can be more than twenty degrees (F.) cooler in a tree-lined street than in a street without any trees.

Tree frog, Dendrobates tinctorius, lives in a tropical forest which provides a moist atmosphere. This species secretes a poisonous mucus from its skin; the skin has a warning type of coloration.

Frog eggs laid by a tropical tree frog on a palm leaf develop in the very humid atmosphere. In temperate regions, frog eggs develop only in water. (Photos made in Panama by Ralph Buchsbaum.)

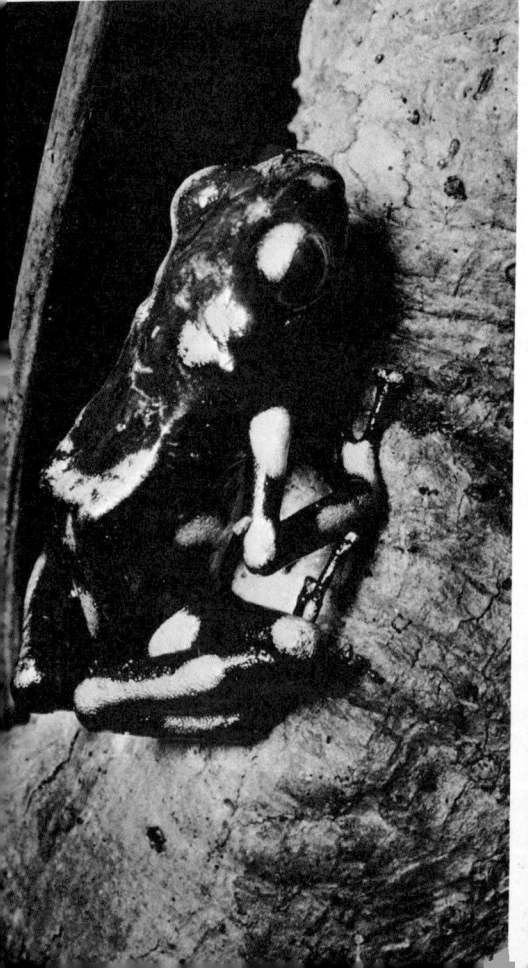

Birds build nests in trees, and many insects breed in rotten logs and under the bark of trees. Such animals are restricted to forested areas. Elf owls live in abandoned holes made by woodpeckers in giant cacti. These little owls are limited in their distribution not only by the giant cacti, but also by the work of the woodpecker. Some mammals hibernate in hollow logs. And man uses trees not only for building houses, but for so many thousands of other things that one leading forester has proposed the per capita wood consumption as the most accurate measure of standard of living all over the world.

Plants are not always passive and exploited by animals. Some plants capture animals and digest them. Insectivorous plants may capture insects, thereby getting an additional source of nitrogen. It is thought that this feature enables them to live in bogs where the nitrogen supply is low. Plants may also parasitize animals. Most of us are, unfortunately, already familiar with "athletes' foot" a kind of fungus which lives in the superficial layers of the human skin. There are many other fungi which grow in man: some on the tongue, others in the lungs, etc. Deriving food from living animals is, however, not a common plant habit. Most plants do well enough by making their own food from soil and air constituents. And some of these constituents, especially the critical nitrogen, are derived from the chemical breakdown, by microorganisms in the soil, of the bodies of dead animals and plants.

Some plant-animal relationships are mutualisms. Certain flatworms (Convoluta roscoffensis) have algae in their tissues. The benefit to the animal is one of added food supply; the adaptation to the algae has gone so far that this flatworm does not feed as an adult. The algae receive a good source of nitrogen and carbon dioxide, and they are transported up and down the shore as the worms migrate with the tide, thus exposing the algae favorably to the sunlight.

Convoluta roscoffensis. Actual length of worms, 1/10 inch. (Roscoff, France. R. Buchsbaum)

One of the best examples of plant-animal mutualisms is that of

the yucca with a moth in the Southwestern States. The female yucca
moth takes a load of pollen from the ripe anthers of one flower to
the pistil of another flower, placing it directly on the stigmatic
surface of the pistil. Then she lays her fertilized eggs in the un-
developed seed pod. The numerous seeds are separated by parti-
tions in the seed pods, and the growing moth larvas seldom eat all
of the seeds. So the yucca is assured of cross-fertilization and the
moth is assured a well-protected and nourished young stage. The
detailed mechanisms in this relationship are an example of mutu-
ally-adaptive evolution on the part of both moth and yucca.

 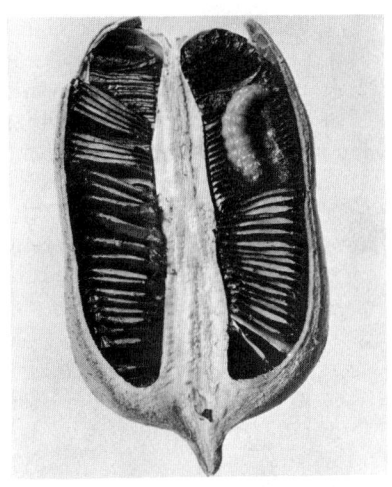

Yucca moth female does not feed, but
lays egg in developing ovary and deposits pollen on the stigma.

Larva of yucca moth in seed pod of
yucca. Larva pupates in soil at base
of the plant. (Photos by C. Clark.)

    Among animal-animal relationships, the most conspicuous is
that of predator-prey. Animals that catch prey and eat meat, the
carnivores, have many fine adjustments of structure that go with
these habits. The sharp, long teeth of a wildcat, a jaguar, or a
lion are efficient for tearing flesh; these are very different from
the broad flattened molars of a cow or elephant used for grinding
grass. For his mixed dietary habits, man has both tearing and
grinding teeth. The ant-eater is a predatory mammal that has lost
teeth in favor of a highly specialized device, an extremely long,
narrow, and sticky tongue that probes rapidly through tunnels in
termite nests. Anteaters also have sharp strong claws for tearing
apart the nests, and their coarse dense fur keeps them from being

bitten by the termite soldiers. Birds have no teeth, but the large and powerful beak of a hawk is a very effective substitute, and the talons of a hawk are truly formidable weapons against herbivores like mice or rabbits.

As was mentioned earlier, two animal species may live together in a one-sided relationship called commensalism, in which one derives benefit, while the other suffers very little or no harm. The echiuroid worm, Urechis caupo, whose specific name means

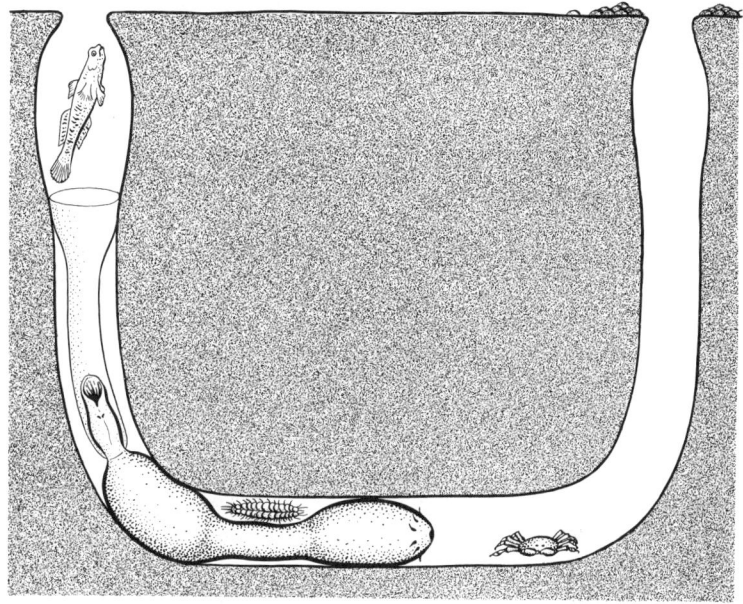

Urechis caupo, the "innkeeper," an echiuroid worm with its commensals: an annelid worm, a crab, and a fish. (Based on Fisher and MacGinitie) x½

the "innkeeper," lives in a U-shaped burrow in certain mud flats on the California coast. In almost every Urechis burrow one may find one or more small fish, an annelid worm, and a crab or two— all of them commensals, like our own household "pets": the dog, cat, mouse, parakeet, or cockroach. They cost their host little or nothing to feed, and they interfere not too much with his activities. And whenever danger threatens, the worm and crab huddle close to their host. In Japan, a newly wedded couple is sometimes given a dried, deep sea, glass sponge skeleton containing a dried pair of shrimps. The Japanese name for this commensal group means "together unto old age and into the same grave." The larvas

of the shrimp enter the central cavity of the sponge when they are
very tiny. They feed upon surplus food organisms brought into the
sponge. As the pair grow to a certain size they become unable to
escape, though their reproductive products can leave in the out-
going currents of the sponge to carry on the species. The first
pair that succeeds in becoming well established prevents the survival of later comers.

Commensalism is probably the starting-point from which most
parasitic relationships have evolved. Something has already been

Glass sponge skeleton and its commensals, a pair of shrimps. Dried
specimen, x½. (Photo by Ralph Buchsbaum)

Hermit crab, Eupagurus bernhardus, has accepted a glass model of a snail shell for its house. This reveals the behavior of its annelid commensal (Nereis fucata), which here (above) at rest occupies the smallest coils of the shell away from its host. As soon as the crab begins to eat (below), the commensal comes to the opening of the shell and snatches a morsel from its host. x½ Marine Biological Laboratory, Plymouth, England. (Photos by Ralph Buchsbaum.)

said about one-sided relationships between plants and other plants
and between plants and animals. Animals parasitize other animals.
Many parasites are so well adapted to their hosts that they have
lost nearly all their sense organs and all means of locomotion in
the adult form. The tapeworm, for example, has lost so much
that it is not much more than a bag of reproductive organs. How
then can the parasite transfer to a new host when the old host is
about to die? Many animals in nature die a violent death—death by
being eaten. Parasites have made the best of what seems to us a
sad fate, by using it as a convenient means to transfer to a new
host. The transfer is effective, of course, only if the particular
part of the host in which a parasite resides is actually ingested,
and provided the predator is a favorable host. Parasites have
evolved very complex life histories which meet this last difficulty.
For example, the tapeworm which lives in the intestine of the fox
gives off its eggs into the intestinal contents. The eggs are shed
onto the soil and get on vegetation. Since foxes do not eat grass,
the young larva gets into another fox by an indirect route. A rabbit eats grass that bears tapeworm eggs; the larvas that hatch
bore their way through the wall of the rabbit's intestine and get
into its muscles and other organs, where they become encysted.
When the rabbit is eaten by a fox, the cysts in the rabbit tissues
reach the intestine of the fox and there become transformed into
young tapeworms.

If both host and commensal or parasite become mutually dependent upon one another, an animal-animal mutualism results. Termites eat wood and chew it up into small particles which they
swallow but cannot digest without the aid of particular flagellates
which they harbor in their gut. The termites use most of the sugar
from this digestion, but the flagellates get a steady supply of wood
particles and a moist safe place to live. Transfer of flagellates to
young termites is effected when nymphs feed upon the fecal pellets
of older termites. If termites are deprived of their flagellates (by
being heated slightly) they starve to death even though they continue
to eat wood.

Another example of animal-animal mutualism is the ant-aphid
relationship in which aphids are transported carefully about to
good feeding places; the ants obtain drops of sugary secretions
from the aphids. This reminds us of man and his domesticated,
milk-giving cows; aphids are sometimes spoken of as "ant cows."

Ants tending aphids in the ant nest in early spring before taking them to "pasture" on the roots or stems of plants. When an ant strokes an aphid, it gives off a droplet of clear fluid rich in sugar which the ant drinks. Most "dairying ants" feed on other things besides the sugar "milk" from their six-legged "cows." (Illinois. Photo by Ralph Buchsbaum.)

RELATIONSHIPS WITHIN A SPECIES

Competition among plants of various species and within the same species was pointed out at the beginning of this chapter. Even the most primitive plants, like blue-green algae, can vie for a place in the sun; competition is a universal relationship, common to all plants and all animals. Though it is individual animals that starve or drown or freeze, the fitness achieved by natural selection is that of the whole population that constitutes a species. A mother bird that dies while defending her young, the male spider that is eaten by the female after mating, the social insects that are unable to feed themselves and must always be fed by others, the female worms that die in the process of shedding their eggs—none of these have evolved toward individual fitness in a world in which they win out by defeating other members of their own species in a world of tooth and claw. The fitness here is of the species as a whole, not of the the individual member. And in the long evolutionary history of the animal kingdom there has been a trend towards minimizing competition within the species, and a selection in favor

of those species which have achieved some degree of internal cooperation.

Cooperation within the species is widespread from the protozoans onward, but it does require increasing levels of structural
complexity to achieve successively higher levels of social cooperation.

Sexual contact may be the briefest kind of cooperation between
two members of a species; for many species it occurs only once in
a lifetime. But the sensory structures and nervous mechanisms
used to bring the two sexes together for mating can also be adapted
for other kinds of cooperation. Mating reactions occur in all phyla
of animals, including many protozoans such as Paramecium.

Aggregations are usually collections of many individuals of the
same species though sometimes they include two or more species.
Aggregations may be formed passively, as when storm waves pile
up mounds of some kind of animal after a storm. Or they may result from the hatching of a group of young animals from a batch of
eggs laid in one spot. Other aggregations are formed by the active
efforts of the individuals as they react to some factor in the environment which is of limited occurrence. Such are the collections
of moths about a light at night, of centipedes under a log, or of
barnacles on a well situated rock. Such animals have achieved the

Passive aggregation of millions of *Velella lata* results when strong winds blow
on-shore during a "plague" of these floating coelenterates. Such huge aggregations occur every few years on Oregon beaches. (Photo made in 1954 by Mrs. G.
Doar.)

level of <u>social toleration</u>. The members do not aid each other,but at least they do not react so as to keep other members of the species from sharing the same favorable conditions.

A higher level of aggregation is that of <u>active positive response</u> to the presence of other members of the species. In this category are many groupings such as those of bees, wasps, crows, robins, and blackbirds that come together only at night to sleep and then disperse in the daytime.

More closely knit are aggregations of birds formed for migrating southward or the many breeding aggregations of birds and mammals. On the other hand, many birds aggregate on common feeding grounds but defend their breeding sites against other members of their own species.

The precision of movement of flying geese or of schooling fish requires both good visual sense organs and a central nervous system capable of coordinating the individual's movements with those of other members of the group. So it is not surprising to find that

Barn swallows gathering in the fall for their southward migration. (Linesville, Pa. Photo by R.B.)

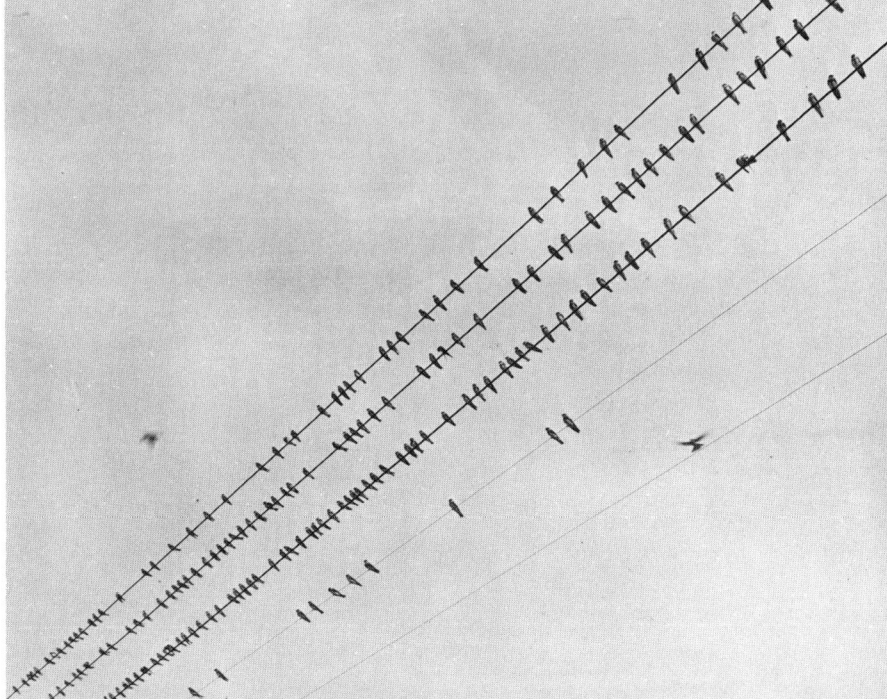

Gannets come in from the sea to aggregate at the breeding season on restricted sites. This is a portion of a colony of 8500 pairs on an island off the Pembrokeshire coast of England. (Photo by Eric Hosking.)

Schooling mackerel have no leader but keep together by reacting to the visual stimuli of each other's movements. (Plymouth Aquarium, England. Photo by R.B.)

increasing levels of social integration do, in general, parallel increasing structural complexity. The arthropod line of structural evolution reaches both its structural peak and its most complex social aggregations in the insects. In termite and in ant societies, particularly, the roles occupied by different individuals are determined by structural differences, and the cooperation of all the members based on behavior that is primarily unlearned and fixed mostly by heredity.

In the vertebrate line of evolution, learned behavior plays an increasingly important role at each higher level of social integration. Mice are loosely organized into colonies with a social hierarchy in which one animal is dominant over several others. But the hierarchy is not present at birth; it is established by fighting. Once the winner of a fight establishes his dominance, the subordinated individuals do not need to be attacked again; they yield to threat alone. So the social hierarchy becomes a means of minimizing conflict in animals that live together. Similar hierarchies exist among many herds of cattle where the recognized leader is usually an old cow (when no bull is present). Many farmers are less aware of the hierarchies among their chickens, where individuals are often harder to distinguish and consistent social relationships not as easy to detect.

In the laboratory, chickens can be marked for recognition, and it soon becomes apparent that the pecking of chickens does not go on at random. The ranking of hens, as in mice, is established by fighting; and once the conflict is over a mere peck or even the threat of a peck will cause a subordinate hen to yield her place at the feeding dish. Any given hen will submit to pecking by a dominant bird. The hen that pecks another without receiving any protest is said to have a "peck right" over the submitting bird. And the social hierarchy in birds is said to be organized by the "peck order." Any bird can rise in the peck order, but only by a successful fight. The resemblance to certain aspects of human societal organization needs no comment.

Wasp nest is raided by army ants while a helpless wasp hovers nearby. Both wasps and ants are highly developed insect societies, though ants have more castes, and the castes are more differentiated in structure and in the work they can do.

Army ant bivouac, *below,* is a temporary nest, complete with chambers, built of the bodies of the living ants. In both wasps and ants cooperation of members is based mostly on unlearned behavior. (Photos made in Panama forest by R. B.)

Fighting mice confined to a small space, where escape is not possible (as in this scene) assume characteristic postures. The black winner, *left*, assumes a dominant stance. The beaten white mouse, *right*, gives the "submission reaction" by rearing up on the hind feet, holding out the front feet towards the aggressor, and remaining motionless until attacked, when it may jump or squeak. (Photo by R.B. of mice studied by Benson Ginsburg.)

Peck-order in chickens. Letters, with A at top of hierarchy, indicate rank order. **A** has just driven **E** into a corner. While **A** is away from feeding dish, **B** has pecked **C** and threatened to peck **D**. (Chickens studied by Allee and students. Photo, R.B.)

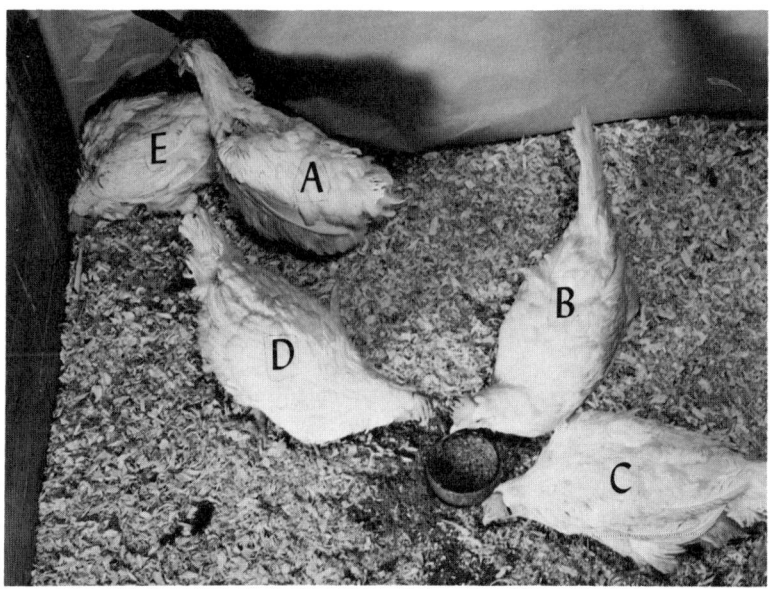

Chapter 4

THE COMMUNITY

A NYONE who can tell the difference between an oak woods and a sandy beach, between a pine forest and a grassy meadow, has made a start towards understanding what the ecologist is talking about when he speaks of biotic communities. The habitable areas of the world are obviously not all alike in topography or in climate, nor are they completely random collections of plants and animals trying to make their way in the world as best they can. We would not expect to find a polar bear in a Pennsylvania woods, nor a starfish in Lake Erie. The variety of physical habitats, interacting with the plants and animals that have come to live in them, have produced definite and characteristic assemblages of plants and animals which we call communities and which include all the living members of the ecosystem.

A community such as an oak woods is not just a collection of trees, squirrels, and owls that continue to live together in the same place because they require similar physical conditions. Nor is it merely a group of plants and animals that live in the same area because they are able to tolerate each other's presence. The importance of the biotic community is that the plants and animals that compose it are actually dependent upon each other and in most cases could not live without each other. If the oak trees were to die, the squirrels and other small rodents would have to leave or to die, not only because their source of food was gone but because the physical conditions of the ecosystem would become too harsh without the tempering influence of the trees that receive the full impact of the natural climate and moderate it, to the advantage of the smaller plants and animals that live in the woods.

In the absence of rodents on which to feed, the owls would migrate. Or suppose only that the owls were all to die of a disease. With fewer predators in the forest, the rodents might increase so as to seriously damage the trees. In a living community there are no heroes and no villains. The predators are as essential to the community as are the less rapacious members. So we may define a biotic community as an interdependent group of plants and animals living in a particular habitat or in a restricted area.

All biological terms which are taken from ordinary language and given a specific technical meaning tend to create some confusion. When we speak of a "nice community" we usually mean only to refer to the members of our own species who live there, not necessarily to the trees, dogs, mice, and centipedes that share the area. But the word "community" has difficulties even when we use the ecological definition. Some physical habitats like islands or lakes end rather abruptly, and their composition, in terms of plant and animal species, differs sharply in immediately adjoining areas. Other habitats grade off almost imperceptibly into adjacent areas, and the biotic communities overlap considerably. This causes no trouble at all for the plants and animals who live in the transition zone (ecotone), apparently because they find it quite suitable for their demands, but it does raise problems for ecologists who are trying to describe and to analyze communities and who find that nature does not always fit into neat categories.

In ordinary speech we use the word "community" to apply to a relatively small group like a "university community" or to the people of two continents as in the "Atlantic Community." Similarly, in analyzing ecological communities we have come to give the term not only to a major community, (the living members of a large forest ecosystem covering thousands of square miles) which is completely self-contained as long as it receives energy from the sun and precipitation from the atmosphere, but also to a micro-community, such as that of a single rotting log.

In a rotting log the microorganisms, beetles, worms, ants, termites, cockroaches, centipedes, collembolas, and others within the log are an integrated and interdependent group of organisms living within a restricted habitat that for a time at least, has relatively little to do with the other animals of the forest. But it is not really self-contained; the energy stored in the wood, which supports the community, was originally stored by the living tree. And in the end the log again becomes a part of the forest floor; it is not a lasting community. Likewise, in an oak woods, there may be a depression which fills with water every spring to form a temporary pond. In this small area no oak trees grow, the sun streams in the year around, the pond edge is invaded by cattails and pond weeds. When the pond is full, in the spring, it supports various

aquatic insects, mosquito larvas, a few frogs, some crustaceans,
a variety of snails, and other small animals that have means for
surviving the period during the summer when the pond dries completely. These animals of the pond, living in the midst of an oak
woods, do have relationships with the forest, as when tree leaves
blow into the pond and decay, when forest birds fly over the pond
and contribute droppings, when the adult mosquito or the matured
frog emerges from the pond and goes to live in the forest. But ecologists find it most convenient to study the pond as a separate community.

In many communities there is some one organism that has much
more mass than any other, as the oak trees in an oak forest, or
the grass in a prairie, and this organism plays the dominant role
in the life of the community. The oak trees not only provide the
protected climate that all the forest animals require, but afford
them shelter as well as furnishing the largest single source of
food. The oak tree is the "dominant" organism of its community
and so we call the community an "oak forest community," if the
forest is a mixture of oak and hickory trees we call it an "oak-
hickory forest community." Others easily named by their dominant
organisms are the beech-maple forest community, or the tall
grass prairie community. Sometimes, especially in the tropics,
there is such a variety of huge trees that no one tree, two, or
even three trees can be said to be dominant and we name the communities by vegetation type, as the "tropical rain forest." As
these samples indicate, the dominant organism is usually a plant
or more than one plant. But there are communities in which the
organism that is most massive and that affords shelter to the others
and also some food, is an animal. An example is the coral reef
community. However, it is not completely self-supporting; the
food supply comes partly from plants and animals that swim or
drift within the reach of the community from the surrounding
waters.

At the ocean we often use physiographic features to designate a
community and speak of the "mud flat community" or the "rocky
shore community." In a river we may distinguish the "rapids community" from the "pools community," where the water flows slowly
and the set of animals is suited to the more static water.

One of the difficulties in naming and classifying communities is
that animals are responding to all of the interacting factors of a

very complex natural environment, and are successful to the degree that they are able to modify their responses with changing
conditions, to employ alternate modes of feeding according to the
vagaries of the weather or the seasonal changes, to move out into
other communities when food runs out in their own—in short, to
respond to their environment with the maximum flexibility that
their biological endowment permits. Ecologists, on the other hand,
are trying to generalize and to simplify, to pigeonhole plants and
animals so that they can be fitted into an ecological structure that
can easily be described and studied.

Though the communities themselves, because they overlap and
because they change, have not yet been completely pinned down,
the community concept is clear enough. And it is one of the most
significant of all ecological concepts, occupying with the ecosystem
concept, the same central and unifying role in ecology as does the
concept of organic evolution in the whole of biology. It brings together into a meaningful story the myriads of detail about plant and
animal structure, distribution, and habits that have been accumu-
lated through the centuries by hunters, farmers, amateur naturalists, as well as by professional scientists. The wholeness of the
ecosystem and the interdependence of the members of the community are simple ideas once they are pointed out. Yet they were
not clearly apparent even to the most careful scholars before the
necessary facts became known.
    One of the most interesting aspects of the community is that it
has a definite structure, and that the structure in its general out-
lines is similar for all communities, including purely human communities. In a human community we recognize certain standard
jobs. These are represented in a rough way by such titles as:
farmer, miller, wholesale grocer, retail merchant, butcher, bus
driver, banker, stenographer, soldier, legislator, preacher,
doctor, teacher, student, fireman, beggar, thief, and so on.
Whether we approve of them or not, each one of these kinds of
individuals has a definite place or job in the community and a more
or less predictable type of behavior, so that if we go around the
world we meet much the same types and expect certain reactions
from them even though from country to country they wear different
clothes and differ in other details. In travelling abroad a teacher
feels immediately "at home" when meeting a French or Italian
teacher but timid about approaching a French farmer or an Italian
bookkeeper. It is the function of the social sciences, which deal
with human ecology, to help us understand the organization of the

individuals in a human community.

In natural communities, too, there are standard "jobs" which
are filled by various individuals; the ecologist calls each of these
a niche. Different natural niches are occupied by different species,
in contrast to human communities, where different niches are filled by individuals of the same species. Hence, a natural community
is much more complex than a human community. We shall review
very briefly here only a few of the common and more important
niches in a plant-animal (biotic) community. Then we shall go on
to show how they all fit together to form a community structure
that is remarkably similar for communities the world over.

1. The food-producers, or basic source of all the food in a
community,are the green plants. These absorb the sun's energy
and use it, together with water and carbon dioxide and minerals to
manufacture carbohydrates, fats, proteins, vitamins, and other
substances which provide energy and building materials to the
dependent forms. Different plants occupy different sub-niches
according to the way in which they do their job and they differ in
their importance as food-producers for other organisms. The one-
celled green alga, Pleurococcus, growing on the north sides of
trees, is a rather unimportant source of energy for the whole community. On the other hand, the plant plankton (one-celled plants
that float in all the surface waters of the seas and lakes) forms
the basic food supply that supports the life of these waters. In
spite of the fact that they are microscopic they support a mass of
animals that is greater than that of the land. The large and conspicuous seaweeds that line the sea shores play a local role as
food-producers, but they are insignificant in the total economy of
the sea.

The plants that convert the most energy, that receive the full
impact of the climate, and modify it for the others, we have already referred to as the dominants of the community. Dominants
to a great extent determine what the other organisms shall be like,
because they moderate the extremes to which an area may go in
respect to temperature, or wind, light, or moisture variations,
and because they constitute the major source of food.

2. The food-consumers. All animals and some parasitic plants
are food-consumers, either of plants or of animals, since they cannot manufacture food from inorganic materials, as green plants do.

2a. The plant-eaters (herbivores) may be small like plant lice
and very numerous, or they may be large like zebras. They may
be specific in their choice of food like certain kinds of caterpillars
that feed only on one species of plant; or they may be generalized
like rabbits or goats. As we see them, their job is to turn over a
large amount of vegetation and to convert it into animal protoplasm—
though, of course, the animals themselves are not conscious of
their "role." The most numerous of the herbivores Elton has
called key-industry animals because they support, directly or indirectly, all of the predators in the community. They are the most
numerous of all the animals because it takes many key-
industry animals to keep one carnivore going. The key-industry
animals must not only reproduce fast enough to replace themselves,
but must make enough new individuals to maintain the species in
spite of the "tribute" exacted by all of the predators which feed
upon them. If they reproduce at a rate greater than that of their
attrition, the species survives; if not, the species becomes extinct.
This is a simple example of the operation of natural selection.
There is nothing of "purpose" in it in any sense of the word. On
land the main herbivores are chewing and sucking insects, field
mice and other small rodents, and the large herbivores like the
ungulates.

In the sea the most important of the herbivores are the small
crustaceans (copepods) which everywhere feed on the floating
plankton and so convert it into a size of organism large enough to
support small fishes. Because of the general absence of large
marine plants, the sea has no large herbivores comparable with
the great herds of browsing ungulates on land. Nevertheless, the
greater surface of the oceans (three times that of the land surface)
traps a comparable amount of the sun's energy. At any one moment
the weight of oceanic plants is only 1000th that of land plants.

2b. First level carnivores that feed on key-industry animals
vary in size depending upon the size of the key-industry animals.
Ladybird beetles feed on plant lice; small birds on earthworms;
coyotes on rabbits; lions on antelopes. These first level carnivores
are generally larger and stronger than their prey and, as previously mentioned, not as numerous. Whereas the key-industry
herbivores may be gregarious, carnivores are usually solitary in
habit.

2c. Second level carnivores feed on primary carnivores and are

larger and fiercer than their prey; they are fewer in number.
Ladybird beetles are eaten by small birds. Another carnivore that
preys on carnivores is the snake which feeds on frogs and toads,
which feed on insects and worms. On every continent there is the
specialized niche of the snake that eats other snakes. In North
America the king snake has this niche. Of course, this third level
of carnivores must contain still fewer members and, again, they
must be still fiercer in some way.

This list does not exhaust the food relations in a community
because organic matter living or dead, in whatever form it may
occur, contains energy-yielding substances; and in the struggle
for existence there will be some animals that find it edible. Hence,
there are other important niches.

2d. Parasites derive their energy by exploiting other living organisms but without killing them. Almost every single individual
plant and animal may be parasitized; rare is the creature that is
not host to a number of other kinds. Man living in different parts
of the world is infected by different kinds of parasites. The same
is true of other species of plants and animals.

2e. Scavengers are animals that eat dead things, plant or animal.
Plant material is salvaged by many animals. In moist soil, earthworms do a great deal of this; in dry soil and in moist soil, ants
are important land scavengers, carrying away dead organic matter of all kinds. Along the shores of oceans, the remains of dead
animals are cleaned up by hermit crabs. In forests, termites and
wood-boring beetles assist in degrading wood and so putting this
material back into circulation in the organic world.

3. Decomposers are mostly microorganisms (particularly bacteria, yeasts, molds and other fungi). They break down the bodies
and excreta of organisms into simpler substances which can then
be converted into nitrogen compounds usable by plants for building more complex substances such as proteins. The microorganisms which do this conversion have been called transformers to
distinguish their niche from that of the decomposers. Together,
decomposers and transformers put back into a form suitable for
plant use the limited supplies of nitrogen available in the world.
The work of the decomposers also releases from dead plant and
animal tissues carbon dioxide, phosphates, and other substances
which can then be reused by the plants.

Meadow in northern Pennsylvania, like any other typical community, has all of the typical niches. The plants, the food producers, here are grasses and various herbs. This ecology student is collecting insects, which are feeding on the plants and on each other. (Photo by R. Buchsbaum.)

One herbivore, the cricket, is abundant in the meadow.
A second-level carnivore, the garter snake eats frogs, toads, mice, etc.
(Photos of cricket, frog, snake by R.B.;

A first-level carnivore, the leopard frog feeds on crickets and other insects, etc.
A third-level carnivore, the hawk, preys on snakes, mice, etc.
hawk by Penn. Game Com.)

Some niches are basic, occurring in every major community and performing an indispensable function, such as that of an abundant green plant or a key-industry herbivore. Others are widespread but limited to certain physical conditions. Earthworms, for example, carry on their scavenging and soil burrowing in moist soils all over the world. Their activities mix plant humus into the upper layers of the soil, and also loosen and aerate the soil so that it is more permeable to air and water. But earthworms cannot live in dry soils like those of our Southwest. There burrowing ants take over the scavenging and soil-loosening niche.

Some niches are highly specialized, yet are widespread. One is that of birds which perform "valet service" for larger animals. The crocodile-bird of Egypt alights on the huge reptiles to devour their external parasites and even enters their mouths to catch flies. Two kinds of African birds brave the irritable rhinoceros to remove bots and ticks. And they do the same for camels and cattle. In Central and South America sun bitterns attend tapirs. And on the Galapagos Islands a kind of land-crab picks ticks off the scaly skin of a giant aquatic lizard. In Africa egrets swarm over the bodies of elephants, catching the insects stirred up by the big beasts. They attend water buffalos in the same way, and cattle also. In America egrets and other small herons associate with cattle both to pick ticks and other parasites off their bodies and to catch the insects stirred up by their hoofs. Occasionally they attend hogs.

Various kinds of other birds frequent domestic animals in the same way. In India the myna bird occupies this niche. In England the starling picks ticks off sheep and deer. In Africa a starling associates with the water buffalo. In North America the cowbirds used to move with the great herds of bison but now settle for domestic stock. And in Panama we have seen the Ani (a kind of black cuckoo) picking ticks off cattle.

No large niche is monopolized by any one kind of animal. The ladybird beetle that specializes on sap-sucking aphids has to compete for them with other small predatory carnivores like spiders and aphis lions (larvas of lace-wing flies). On the other hand, animals that feed in the same way in the same general area do not necessarily occupy the same niche in competition with each other. Catching insects on the wing is a way of life for the swift (a bird), as it is for the dragonfly and for the bat. But the swift can feed in the daytime, the dragonflies mostly at dusk, the bat during the night. The dragonfly hunts over the borders and open water of ponds and streams, so it does not necessarily compete with a nighthawk that patrols meadows in search of flying insects at dusk. Even among the many species of dragonflies, smaller species feed at lower levels than do the larger species.

A question often asked is whether related species in the same area usually occupy the same or different niches? The answer is, with rare exceptions, different niches. This is because a group of animals that continues to occupy exactly the same niche does not usually break up into separate species. The very act of occupying a new niche provides the isolation which favors species formation. There are many good examples of this at the ocean shore, where the meeting of land and water creates niches with very different conditions, sometimes within a few yards of each other. On the sandy beaches of Normandy and Brittany in France, in the moist sand between tidemarks, one sees the large green patches that turn out to be, on close examination, huge aggregations of Convoluta roscoffensis, a tiny green flatworm that harbors green algae in its tissues (as described on page 45). The worms migrate up and down with the tides but are limited seaward by the depth at which the worms, though partially covered with water, still receive a maximum of light for the photosynthesizing algae, and limited landward by the highest reach of the water at the least extreme of the high tides. A short distance seaward, in the zone never uncovered by any tides except the lowest ones of the year, lives Convoluta paradoxa, another species of the same genus. It has different habits and lives on the rock-attached seaweeds.

Certain groups of animals, through millions of years of natural selection, have developed the structure and habits that enable them to take over and hold on to a particular kind of niche in any part of the world into which they have been able to penetrate. Sometimes their niches are unoccupied when they arrive, often they must displace a less well-adapted earlier arrival. In modern times we have repeatedly seen alien placental mammals introduced into Australia displace their ecological counterparts when the two are thrown into direct competition. Australia is an isolated continent difficult for mammals of any kind to reach. Apparently the marsupials got there first and spread out into many niches which on other continents came to be filled by placental mammals. Now that man has brought placental mammals with him to Australia, many

Odd niches may be found in every kind of community; a few are shown here.

1. Pitcher plant, growing in a marsh. traps insects in its pitcher-like leaf and digests them, thereby adding to the meager supply of nitrogen from swamp water.

2. Ant-eater, in Panama, is specialized for tearing apart termite nests and probing the tunnels for termites with its sticky tongue.

3. Spider, in Panama, captures insects in its purse-like web.

4. Hermit crab, a scavenger of the sea shore.

5. Dung beetle, gathering a ball of manure in which the female will lay eggs; this ensures a supply of food for the young.

(Photos: 1-4 by R.B.; 5, C. Clarke)

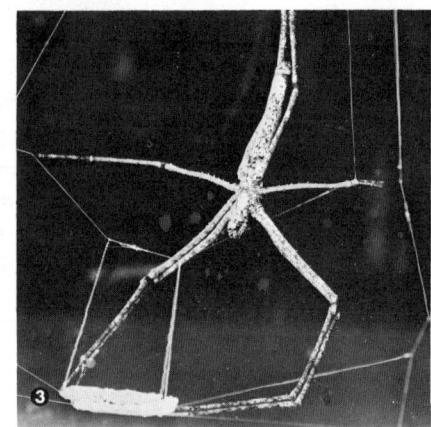

marsupials are being pushed out of their long-held niches by placentals such as rabbits, sheep, cows, and dogs.

Though niche and biological relationship go together more often than not, so that it may take a taxonomic expert to point out to us the species differences that distinguish the barnacles and snails and limpets of our Oregon Coast or our Maine Coast from those of the ecologically similar Cornish Coast of England, there are also many exceptions. Sometimes it takes a taxonomic expert to notice that animals of very similar appearance, and occupying the same niche on different continents, are not as closely related as their appearance and habits suggest. They look alike because they have become adapted, through long natural selection, to the same set of conditions. Even on different continents, animals that fill the same niche resemble each other more closely than we would predict from knowing the taxonomic groups to which they belong. There are particularly good examples among birds, where conspicuous coloration attracts attention readily to superficial resemblances.

The American meadowlark, of the genus Sturnella, has a coloration unusual for the family to which it belongs. The African pipit, of the genus Macronyx, is also unusual in color and markings for its family. Yet the two birds are amazingly similar in appearance, with streaked dark upperparts, and bright yellow underparts on which a broad black band crosses the breast. Both turn aside and hide their brightly colored underparts when approached by an observer, and they also spread the tail when flying, exposing white side feathers. Both have a "melancholy" whistling note. And both inhabit grassy open places (in the U. S. the Western meadowlark is a typical bird of our prairies (see page 111). Meadowlark and African pipit make their nests on the ground and arch it over with dry grasses. And to top off the story, there is another genus, Pezites, in South America, which has the same habits and which closely resembles one of the African species of Macronyx, in which the breast is pinkish red instead of yellow.

Our red-winged blackbird, whose detailed structure puts it into the same family as the meadowlark has, in the male, the same black coloring and red "shoulder epaulets" as a genus of African weaver-birds. The two look-alikes also share the same general habits. The appearance in two species of similar structural features or habits which cannot be accounted for by the common

heredity of close taxonomic relationship is called <u>convergent</u>
<u>evolution.</u>

FOOD CHAINS AND FOOD WEB

Our American community is not a typical human community.
Only one-fourth of the people in it are able to produce most of the
food of the other three-fourths, and these latter earn their living
in some way which is not related to food production. Moreover,
the food we eat comes not from the particular habitat in which each
of us lives, but from every kind of climatic zone and from all parts
of the world. We prepare coffee from the Brazilian tropics, tea
from the Asiatic tropics, import bananas from the Central American tropics, and pineapples from tropical Hawaii. We grow our
wheat and corn in our own grasslands, nuts in temperate forests,
dates in the Southwest, domestic stock on western plains. We eat
whitefish from Lake Superior, salmon from the Pacific Ocean, cod
from the Atlantic, lobsters from both Atlantic and Indian Oceans,
and shrimps from the Gulf of Mexico.

A food chain in the sea. From left to right: plant plankton are eaten by animal
plankton (mostly copepods) which are eaten by small fish; . . .

Once man had to eat mostly foods of a size he could grasp in
his bare hands: fruits, shellfish, small mammals. Now size to us
is irrelevant, and we raise millions upon millions of tiny wheat
seeds, then use machinery to grind them into flour and afterwards
make loaves we can handle conveniently. At the other extreme we
eat meat from steers that no man can lift, and occasionally even
feed on steaks from a whale that weighs as much as 2000 men!

This is very different from the life of people in the older countries like India or China, where resources can no longer be borrow-
ed from the past or surpluses exchanged for foods from afar. Only
4% of the population of India is in industry; most of the rest are
concerned during all waking hours with the production of food.

The Indian situation is more like that of a natural plant-animal
community, where producing and consuming food is the main order

of business and goes on all the time that plants receive sunlight
and animals remain active. Merely to know what an animal eats is
to know a great deal about how it lives and what it contributes to
the life of the community (as we saw in the preceding discussion
of the niche). All members of a community are linked together by
their eating-eaten relationships, so we can make a good start at
understanding the community by following out these linkages, which
we call food chains.

   Food chains follow a general pattern: green plants→herbivores
→carnivores→still larger carnivores, and so on until we come to
a "top carnivore" that has no larger predators. On land, where
there are many large herbivores, food chains often are short
and have only two or three links.   In our western grasslands it
could be: grass→cattle→man; in an African grassland: grass→
zebra→lion. In a Pennsylvania meadow, with small herbivores, there
might be five links: grass→cricket→frog→snake→hawk (as shown
on the front cover). In aquatic habitats herbivores are usually minute, and it usually takes five or more links to convert the plant sub-

. . . these are eaten by larger fish and these, in turn, by still larger predatory
fish.

stance into an animal that has no larger enemies. The sequence in
a pond could be: algae→protozoan→small aquatic insect→large
aquatic insect→black bass→pickerel. In the ocean, chains may
have five links, as in the diagram above, or more.

   The food chain is useful as a working tool in tracing out who
eats what, but by itself it gives an inaccurate picture of what is
happening in a community. Many animals have become structurally specialized and in this way they have been able to exploit a source
of food unavailable to others. They have found a new niche, and
the rewards are comparable with the "bonanza" that may come to
the inventor of a new device or a new service in human society.
But there are serious limitations for the animal specialist. It can
feed in only one way, and sometimes only on one species of plant
or animal. Most members of a community are not so highly specialized and they can eat more than one kind of food and regularly

do so. So-called "seed-eating" birds eat insects in the spring. Foxes eat mice when they are very abundant, go after rabbits when mice are scarce, gorge on berries when berries are ripe, shift to fallen apples and grasshoppers in the fall, and so on (as we saw in the diagram on page 42). Small herrings eat not only copepods but almost any kind of small marine larvas of snails or barnacles. Small herrings are themselves eaten by arrow worms, which are then eaten by larger herrings.

In a pond the clams are specialized to strain microscopic organisms out of the water, and the copepods are too small to eat anything larger than one-celled algae and protozoans. But most of the other animals in the pond eat any animal of a size they can swallow.

Such relationships, in which every predator eats several kinds of food and every kind of food is eaten by many different animals, cannot be expressed by a row of parallel chains lying side by side. When diagrammed, the total of all the food chains in a community becomes a food web.

That the food web has any sort of consistent pattern, instead of being a scramble of accidental encounters that change from one moment to the next, is inherent in the feeding process. Plants (with rare exceptions noted earlier) cannot feed on animals, nor can herbivores eat carnivores. And carnivores cannot devour just any animal they meet because some kinds of prey run away and those big enough to defend themselves will do so. Such matters are no problem for herbivores, because plants cannot run away nor do they fight back. Minute aphids suck the juices of the loftiest trees, and small ants can in a few hours defoliate a large shrub. For carnivores (and omnivores when they eat animals) the size of food is often critical. They must overpower and subdue their prey by sheer strength, and this usually requires superior size.

It is apparent then that the main factor in determining the direction that any predator-chain takes in a food web is the size of animal. Animals below a certain size are uneconomical as a food supply; it may take too long to collect enough of them to fuel an active animal. Animals above a certain size are too difficult to overpower. Lions do not bother with grasshoppers, nor do spiders capture lions.

Both the upper and the lower size limits of a predator's food may be extended either by sheer necessity, or by specialized adaptations, to take advantage of available food. Coyotes ordinarily eat small rodents because they are easy to catch when reasonably abundant. After one of man's rodent-poisoning campaigns, a coyote may be driven by hunger to attack an animal as large as a sheep. By hunting in packs, wolves can kill a mammal much larger than themselves, and the same is true for a band of predacious ants. Animals that inject poisons can also kill prey larger than themselves.

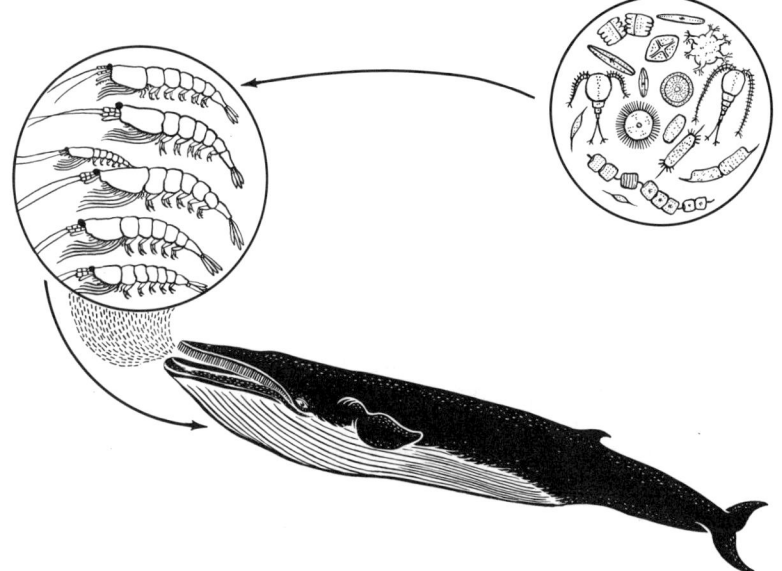

A short food chain: plant plankton and copepods → euphausids → blue whale. (Modified from Popovici and Angelescu.)

Since the surface waters of the sea support so much microscopic life, the sea has a tremendous variety of clams and worms (see Urechis on page 47) and other animals that live by straining and concentrating minute organisms of the waters in which they live. Most remarkable, perhaps, is the 100-ton baleen whale that strains tons of water through its whale-bone apparatus and so is able to support its tremendous bulk on the small shrimp-like crustaceans that occur in almost unbelievable densities in the colder waters of the seas.

Even after we take into account all the carnivores that feed on

food much smaller than themselves, and the relatively few which
by special devices or a ferocious appearance can kill animals of a
superior size, it is still true that all predators (except modern
man with his tools) are for the most part limited to food of a particular size range. The kind of food may vary tremendously, as
we saw for the fox, but the size of food has definite limits. And it
is this size limit that gives an orderly pattern to food chains and
to food webs.

A food web on land that has been worked out fairly completely
is one on a relatively barren island, Bear Island (in the Arctic
Ocean north of Norway). This one is shown here because of the
relatively small number of different kinds of animals involved. In
the tropics or even in the temperate zone, such a diagram would
be confusingly complex.

A marine food web is also given here (on page 77), but it is
limited to the food relationships of the herring. It is incomplete
in that it does not show that the largest herrings are eaten by
sharks, as can be seen in the diagram of the food chain on pages
70 and 71.

In the sea, as on the land, the total energy requirements of the
members of the food web are in dynamic equilibrium, that is, as
much energy enters the system as leaves it. Since all energy
comes from the sun, and green plants alone can convert this energy into a form utilizable by animals, the total amount of animal
life must be less than the total amount of plant life on whose surplus energy the animals depend. Each animal, in its life-activities,
uses up some of the energy contained in the food and dissipates it
in the form of work and heat. Further, the smallest herbivores
are able to multiply very rapidly chiefly because they are so small,
and so produce an excess number of individuals over that required
to maintain the species. This margin provides a living for a group
of carnivores which are larger in size and fewer in numbers than
the herbivores upon which they feed. The total bulk of carnivorous
animal life cannot be as large as that of the herbivorous species.
The total number of crickets (herbivores) in a meadow is enor-
mous; the frogs which prey on them are fewer; the snakes which
eat frogs are still fewer; the hawks are rare. One lion may kill
50 zebras in a year. Although the zebras reproduce slowly, they
are numerous enough to maintain their own numbers while producing the surplus zebra "crop" which supports (in part, because

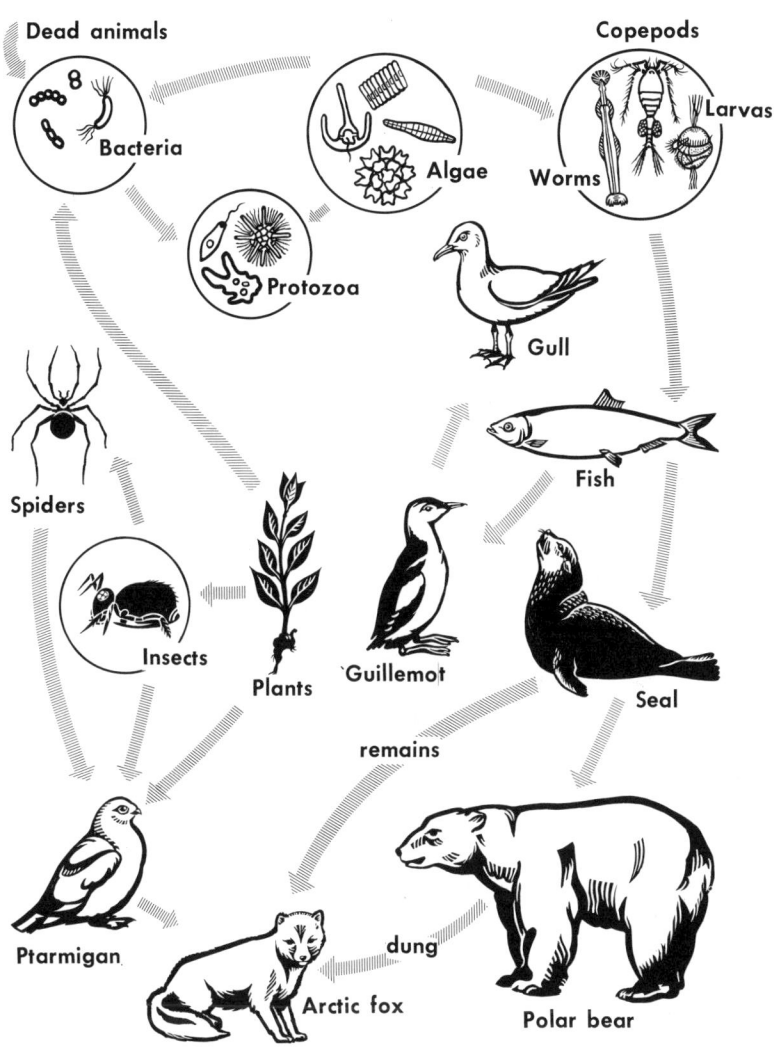

Food web on Bear Island in the Arctic zone. Most of the food relationships are shown, but some are omitted. There are many more species of protozoans, algae, rotifers, copepods, insects, birds, and fish. The arrows are read "eaten by," for example: seal ➔ Polar bear means that seals are eaten by polar bears. (Simplified from Elton.)

lions also eat antelopes) the much smaller lion population. The lion population, however, is too small to support any larger or fiercer carnivore—even if there were such. In a pond, the number of unicellular algae runs into billions; the protozoans which eat them run to millions; the small arthropods into thousands; there are hundreds of minnows, and a few larger fish. It has been calculated that the plants in the plankton of a lake (such as Lake Mendota, Wisconsin) weigh twelve to eighteen times as much as the animals in the plankton. The total weight of the small fish which eat the animal plankton would be much less.

PYRAMID OF NUMBERS

Diagrammatically, the basic relationships may be represented by a pyramid of numbers. It is easy to see that if the amount of energy put into this system is changed at any step along the way, all the other steps would have to adjust to the changed conditions. Most serious would be anything that would affect the base of the pyramid. That is why all studies of the productivity of the sea, for example, center about the plankton instead of about the more exciting animals higher up in the food pyramid.

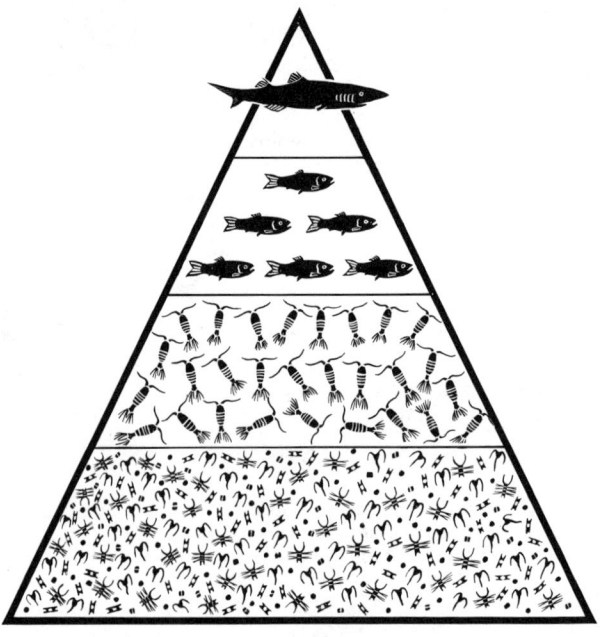

Food pyramid in the sea showing the quantitative relationships not shown by a food chain or a food web.

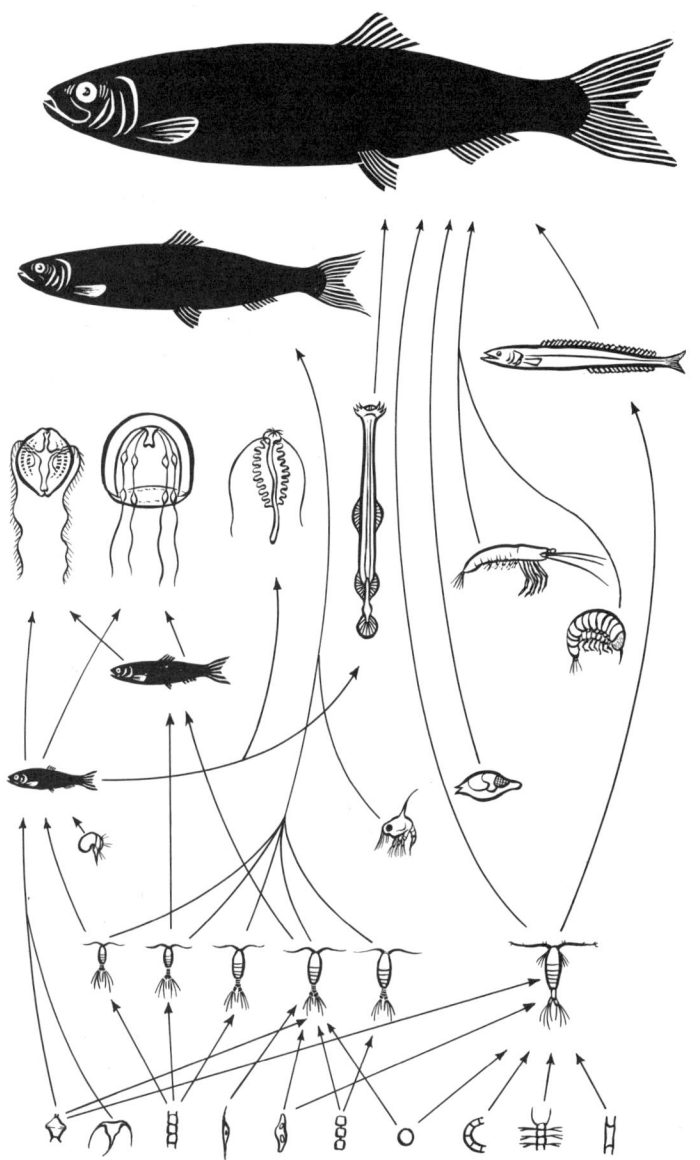

Herring food web. The herrings are shown in solid black. Note that the smallest herrings are themselves eaten by (from left to right) the comb jelly, the jellyfish, the tomopterid worm, and the Sagitta or arrow worm. In the bottom row are the microscopic plant plankton, above them various species of copepods, next higher are the small planktonic larvas of snails, barnacles, and crabs, etc., then the smallest herring ½ to ¼ inch in size, above this a herring ½ to 2 inches, crustaceans about 1 inch in length, above these four larger animals (1 to 3 inches) of the plankton (including the comb jelly, etc.), then the sand eel, next the 2 to 5 inch herring, and at the top the herring 5 inches and over. (Based on data of Hardy.)

It is interesting to consider how much of the sun's energy that strikes the earth finds its way into man's nutrition. During an average growing season, an acre of corn will produce about 35 bushels of grain, or (at 56 pounds per bushel) 1,960 pounds. This is equivalent to about 3 million calories and represents less than 1% of the energy of the sun during the growing season. At about 3,000 calories per man, the acre of corn will support 1,000 men for one day. If the corn is fed to steers, and the meat is eaten by men, only about 5% of the energy of the corn is available for human nutrition; this would amount to about 150,000 calories, or enough for 50 men for one day. In terms of the sun's energy, eating meat results in utilizing not 1% of the sun's energy, but 0.05% (or less) of the sun's energy. If we wish to obtain the most energy from an acre of crops—all other considerations of such things as taste and adequate nutrition aside—we should eat the plant products directly instead of feeding them to animals. In terms of energy, eating cattle is "eating sunshine" third hand. It is wasteful, because each link added to a food chain dissipates some of the energy in carrying on its feeding and other activities.

The top carnivores in any food pyramid, such as hawks, lions or sharks, do not lead an idyllic life just because they are not preyed on by larger forms. Like man, they are in competition with other members of their own species for food, shelter, and mates, and with other carnivores for food. They are limited in number not only by the amount of energy they can capture, but also by the amount of damage done to them by their parasites. All along the line from the smallest to the largest animal in the food chain, parasites occur, so that there can be discovered a parasite chain—a "bigger-fleas-have-lesser-fleas-on-their-backs-to-bite-'em" sort of thing. The larger the animal the more parasites it

Tick, with its beak inserted in human skin, begins to feed.

Engorged tick, of the same kind, compared with another which has not fed. x2 (R.B.)

can support. Lion or man may have ticks on his skin, protozoans in his blood and intestine, flukes (flatworms) in his nose, blood, or intestine, and tapeworms in his intestine. Moths may lay eggs in the lion's fur, and when the larvas hatch, he will become moth-eaten. Both lion and man may have several kinds of fungus disease. The larger parasites, like the ticks, may have within their bodies parasitic worms and protozoans. And these worms are in turn parasitized by protozoans. Of course, all of them may be invaded by bacteria, which in turn are plagued (as all animals are too) by viruses. It need hardly be pointed out however, that the total bulk of the parasites in a lion or a man must be considerably less than the bulk of their host (see front cover).

Just as the carnivores in a food chain become larger in size and fewer in number, so in a parasite chain the parasites become smaller in size and larger in number in each link of the chain. One dog may support (uncomfortably!) hundreds of fleas; and each flea a great many protozoans. However, the total bulk of the fleas is very much less than that of the dog, and the total bulk of the proto-zoans is less than that of the fleas.

Plant-animal relationships do not always involve direct encoun-ters. All plants and animals are related in the carbon cycle in which plants remove from the air the carbon dioxide expired by animals. Plants produce carbon dioxide in their respiration also, but in relatively smaller amounts. The carbon dioxide taken up by plants is used in synthesizing food, and this process of photosyn-thesis produces two by-products, water and atmospheric oxygen. Oxygen so produced is used by all animals (except those few which liberate energy anaerobically), and also by plants, in the respira-tory process by which they oxidize food substances to release the energy needed to carry on living processes. This biotic relation-ship, on more penetrating analysis, reveals itself to be one that involves purely physical factors.

Most of what has been learned about food chains and food webs and the qualitative aspects of the food pyramid has been accumu-lated by observation of plants and animals in natural communities. The information is still very incomplete, and any ordinary person who makes careful observations in his garden or near a summer cottage can, by keeping careful records, help ecologists to add to the store of facts about what animals have been seen to do and eat.

In addition to simple observation of animals in the act of feeding, ecologists examine the stomachs of many animals, the pellets
of indigestible residues cast up by birds, the refuse in and around
bird nests, the fecal pellets of many mammals, the characteristic
scars left on acorns and nuts by the teeth of rodents, the snow
tracks of mammals stalking their prey, the litter left in burrows
and other animal shelters. But for special kinds of information
on feeding, on migratory habits, and on numbers of individuals,
it is necessary to introduce certain artifacts into the community.

In England, a student of birds, Lack, has banded many English
robins so that he can recognize individual birds living in the wild.
He has been able to tell a great deal about their social behavior
that cannot be seen at all when any robin looks exactly like any
other to the observer. Another ecologically-minded ornithologist
banded mice in an oak woods and then collected the metal leg bands
from the nests of tawny owls that ate the mice. From the markings
on the bands he could tell how much territory in the forest was
patrolled by one tawny owl, since he knew exactly where in the forest he had released the banded mice. It turned out that the territory
monopolized by any tawny owl, as indicated by the evidence from
the banded mice, was exactly the same as the territory he had as-
signed to each owl by listening to its hooting in the dark. So he was
able to verify the long-held suspicion that the hooting of the tawny
owl, like the cheerful songs of many song birds, is a vocal threat
to other birds of the same species that the bird doing the vocalizing
"owns" the territory from which it vocalizes and that it will fight
to defend it. Fighting is seldom necessary. Most birds accept the
ultimatum and keep to their own territories. In this way, injury to
the species is minimized, yet each bird dominates a nesting territory. Territoriality has been considered a device for ensuring adequate feeding ground. But most birds enforce it only in breeding,
while sharing feeding grounds. Perhaps it arose as a way of spacing out nests as a protection against predators, though in some
birds it does serve as an aid in feeding the young.

Estimating the size of wild populations by releasing banded animals and then trapping at random to determine the ratio of banded
to unbanded ones that are caught in the traps, is a common method
of study. And there are dozens of other useful methods of analyzing
natural communities without disturbing the animals too much.

Some things we wish to study, however, cannot be learned easily in the field, where weather and other conditions are too variable. Nor do ecologists care to spend the whole winter in a forest if they can help it! So without minimizing the drawbacks of laboratory study, they set up laboratory populations of certain kinds of small animals and measure the effects of various physical factors by varying only one factor at a time. Or they study the effects of various known degrees of under-population or overcrowding. One of the things they learn about overcrowding in mammals, both in the field and in the laboratory studies, is that it makes rats or muskrats neurotic, and so irritable that they do not grow well or breed well, even when food is plentiful. Yet we always glow with satisfaction when we read that our own city has increased the density of its population, or feel that the city administration has somehow failed when we read that the latest census shows a decrease in population density.

One of the advantages of laboratory study is that it enables us to set up artificially simplified communities which are easier to analyze than the complex natural ones. Such artificial setups also lend themselves more readily to quantitative analysis. One student of protozoan ecology set up a community consisting of only one prey animal, Paramecium caudatum, which was fed on a bacterium, Bacillus pyocyaneus. In this microcosm it was possible to study the action of a single predator, Didinium nasutum, under carefully controlled conditions and to test hypotheses based on purely theoretical and mathematical considerations.

Moreover, it is possible to set up a laboratory relationship between two species that never would live together in nature. Many complex tissues, even of man, will grow and multiply when grown aseptically in glass chambers if kept at the normal body temperature of the animal concerned and if supplied with nutrients. Such a "tissue culture" can be made from the connective tissue of a chicken embryo and then supplied with green algal cells (Chlorella) to form an artificial mutualism. When such a combined culture is grown in the light, both chicken cells and algal cells grow better than when either is grown alone under the same conditions or when the two together are grown in the dark. Presumably, the animal cells benefit from an added source of oxygen, which is in short supply in a closed culture, or are freed from an accumulation of carbon dioxide. The algal cells presumably receive a steady supply of carbon dioxide and of nitrogenous wastes from the animal cells. Such an artificial mutualism can be studied under conditions

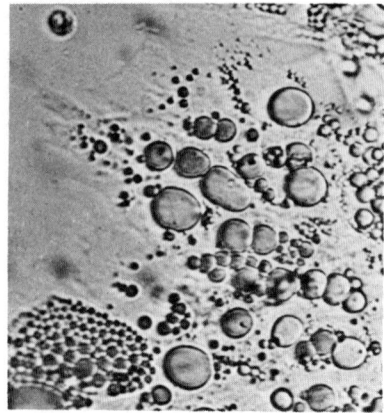 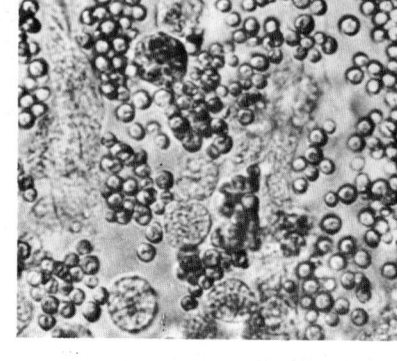

Chick connective-tissue cells grown
without algae in the light for 8 days.
Most of the cells are packed with fat
droplets and are dead. x635

With algae, chick connective-tissue cells
grown in the light for 8 days. Algal and
animal cells are in good condition; fat
droplets are rare. x635 (R. Buchsbaum)

relatively easy to control and to analyze. It has been cited as evidence that many organisms evolve advantageous capacities long
before they actually are of use in nature and that it is such "pre-
adaptation" that enables animals to step at once into some new
niche as soon as it arises. Such adjustment to special conditions
thus may precede the action of natural selection in further per-
fecting the adaptation.

Chapter 5

PERIODIC CHANGES IN COMMUNITIES

IN HUMAN COMMUNITIES we seek to insulate ourselves, in so far as we are able, from the natural variations in our physical environment. In our mechanized homes we turn night into day, winter and summer into perpetual spring. We eat fresh fruits and vegetables the year around, even though, more and more, we may sacrifice both the flavor and the vitamin content that made this desirable in the first place.

How well we have succeeded in our big cities, in bending the environment to our demands, comes to our attention most vividly when we spend some time in a rural or remote place. There we soon may find it necessary to imitate the natives by going to bed with the setting sun, arising with the birds at dawn, putting off any kind of late errand to a night with a full moon, and dressing for the seasons as they come. We eat fresh fruits and vegetables only when they are in season, gorging on strawberries one month, then turning to nothing but blueberries as they ripen. In winter we may eat none but stored fruits and vegetables, in very limited variety. In summer we are up eighteen hours a day, in winter sleep much more and stay indoors for days at a time. In this way of life we come much closer to appreciating the solar and lunar rhythms that affect plants and animals.

Animals adapt to environmental changes in many of the same ways as man, though mostly without his use of artifacts and of learned behavior. They change their body insulation, dig or build shelters, store food from one season to the next, migrate to a warmer community in winter, etc., but achieve these adaptations largely through inherited structural and behaviorial adjustments. As was pointed out in Chapter II, under plant adjustment to the physical environment, plants have almost no behavioral adjustment; they cannot dig burrows, or migrate. They must meet changes in the environment where they are rooted, and do so by some kind of structural adaptation.

Some communities are better insulated from changes in the physical environment than are others. On the ocean floor, at great depths, the darkness is perpetual, the temperature practically constant, and the seasonal animal activity varies little. Certain caves also have a steady sameness of physical conditions and vary only in the amount of food that enters the cave from outside sources or in the coming and going of animals like bats. These are the exceptional communities. Most are subject to all kinds of short-term and long-term changes.

DAY AND NIGHT

During the night hours in a big city, the human fauna includes few faces that would be familiar to a member of the daytime community of the same streets and buildings. A doctor hurrying to deliver a baby or a university student working at the postoffice at night and attending daytime classes might overlap between the two communities. But most of the night workers are almost as isolated from their daytime counterparts as if they were in a different city. Some niches are the same: the night and day shifts of a restaurant, of the postoffice, of the mills, or of a fire station. But most daytime niches in shops and offices stand empty, while special night niches are worked by the crews that clean office buildings and by night watchmen.

During the day, as in the night, activity is low at certain hours, rises to peaks at the same hours of every cycle. In the early morning the milkmen make their rounds before most people are up. Then streets fill as workers go to factories, clerical workers to offices, clerks to shops, and students to their classes. At noon comes another peak when people pour into the streets and, after lunch, as suddenly disappear. At dusk this human fauna overflows the streets as it returns in frantic haste to its shelters. Then in mid-evening there is another flurry as theater goers arrive, followed by night workers and night watchmen—and another night cycle begins.

In natural communities the day and night shifts are worked, for the most part, by different species. When one set goes to sleep the other takes over many of the same niches, but not all of them. Nocturnal animals have none of man's artificial devices for overcoming darkness. They must come structurally equipped to their work in the dark, and some of them do have sensory equipment which diurnal animals like man do not have. The bat (see page 34)

can fly about in pitch blackness without striking obstructions, and
to do this has a kind of "bat sonar."

Day-flying wasp, left, builds nest, collects food, and works during the daytime; it
sleeps at night. Its niche is occupied during the night by a night-flying wasp, right,
which is adapted to night activity; it has larger ocelli (simple eyes, indicated by
arrow) than the ocelli of the day-flying species, and its body has numerous light
yellow markings. This species is inactive on its nest during the daytime. Panama.
(Photo by Ralph Buchsbaum.)

DAY AND NIGHT IN AN OAK-HICKORY WOODS

An oak-hickory woods in summer is about as busy a daytime
community as one can find among temperate forests. The mixture
of shade and open sunny spaces affords more niches than either a
shady beech woods or an open meadow in the full sun. On the forest
floor thrushes are busy pulling up, from their burrows underground, the juicy earthworms that retired when the sun came up.
A thrush that becomes too intent on his feeding falls prey to a hawk,
or even to a red fox that usually prowls in open woodland but comes
deeper into the forest whenever the hunting there is good. In
early summer chipmunks dig about for roots in the floor humus;
in fall they gather acorns and hickory nuts from under the trees.
They often fall prey to the ground-roaming red fox.

In the shrub layer above the forest floor are a multitude of insects of many kinds assuring a food supply for the frogs and small
insect-eating mammals, and even the red fox. Conspicuous in daytime are butterflies and bees that go from flower to flower and are
fed on by birds. Day-biting mosquitoes seek out ecologists.

Overhead the forest is noisy with the tapping of woodpeckers, as they search out ants and grubs from the bark of tree trunks. High in the trees, in the leafy crowns, the tree squirrels gather their plant food, safe at this season from the red foxes.

Then comes dusk, and most of the daytime animals begin to go to sleep. In this transition period between day and night, dragonflies fill the air over the pond margins, taking advantage of the overlapping of day and night forms to catch both diurnal and nocturnal insects on the wing. When darkness settles, the din of the cicadas changes to the chirping of crickets and the shrill songs of the katydids. The tapping of the woodpecker ceases, and its niche, which would require good vision in the dark, stands unfilled. But the call of the whippoorwill tells that it is awake and will be busy catching night-flying mosquitoes and big nocturnal moths. Bats swoop about with mouths wide open, gathering moths and other insects, a mammalian niche which has no counterpart in the daytime. Deer mice take over from chipmunks and squirrels, and are captured by the owls that take over the night shift from the day-flying hawks. Though the tree squirrels are asleep, the flying squirrel is active in the treetops hunting his favorite food, the hickory nut. This gliding squirrel (which cannot really fly) is the most nocturnal of all American mammals. Though difficult for us to see, we recognize its presence by the distinctive "signature" it leaves on hickory nut shells. The gray fox, which lives deeper in the forest than the red fox, hunts mostly in the night. The striped skunk is largely nocturnal, hunts insects and mice, but has to keep an eye out for its enemy the horned owl. The opossum, our only North American marsupial, is arboreal and mostly nocturnal. It is said to be particularly fond of eggs but will settle for almost anything organic from snakes to nuts. The raccoon comes down from his tree at dark, and may bring his whole family along to help catch mice, frogs, and insects. The short-tailed shrew, a small burrowing rodent that eats both insects and rodents (some larger than itself) is said to be primarily nocturnal. But shrews have such high rates of metabolism that they need to eat almost constantly, and they can be seen, ferociously attacking their prey, at any time of the day or night.

Day-night changes in plants are seldom conspicuous externally, but they are more basically thoroughgoing than in animals. Photosynthesis can be carried on only in light, of an intensity which differs with the species (see also page 33). During the night photosynthesis stops and sugars manufactured during the day are trans-

ported downward to storage tissues. A few plants, like the sun-
flower, turn with the sun. Many plants open or close their flowers

Night-blooming Cereus flowers once a year yielding this beautiful and fragrant
white blossom which lasts but through one night. (Bermuda. Photo by Buchsbaum.)

at particular times of the day; some plants open their flowers only
at night, when the insects that pollinate them are active.

DAY AND NIGHT IN THE DESERT

Extreme day-night changes characterize the desert. In the heat
of mid-day, when the sand may reach a temperature of 170°F.,
hardly any animals are abroad. In the early morning, and in the
late afternoon the birds of our southwestern desert feed on insects,
spiders, seeds and berries. The "road-runner," a predatory bird,
swiftly pursues snakes and lizards. The turkey vulture circles
overhead on the lookout for dead animals. Jackrabbits may come
out of their burrows in early morning but retire again as the sun
rises in the sky. The ground squirrels frisk about a little longer,

but they soon seek shelter too. Then just before sundown, all these
animals become especially active as they take their last daytime
meal, and some of them overlap with the early risers of the nocturnal shift, so that dusk is a good time to see both day and night
animals of the desert.

One of the few lizards that comes out at night is the poisonous
Gila monster. The bull snake is now asleep underground, and the
desert rattlesnakes occupy the snake niche. But it is the mammal
that really takes over when most of the lizards and birds retire
for the night. Small rodents, of largely herbivorous habits, come
out of their burrows in great numbers: the deer mouse, the pocket
mouse, the pack rat, the kangaroo rat. Having rested during the
hot hours of the day, the black-tailed jackrabbit is active again,
filling the rabbit-sized herbivorous niche. The rabbit-sized omnivorous niche is filled by the small spotted skunk, which feeds on
cactus fruits, insects, and baby mice and rabbits. Like its larger
temperate forest relative, it falls prey to the horned owl. Finally,
there are the larger mammals, all of which hunt rodents. The
ring-tailed cat goes after pack rats; the kit fox after kangaroo rats,
and in turn falls prey to the bobcat. Larger than any of these predators is the coyote, but it has often to be satisfied with mice or
even with carrion (flesh of decaying animals) and with mesquite
beans. Well adapted to the niche of the large-sized digging mammal is the badger. It uses its strong front feet, armed with sharp
claws, to dig sleeping diurnal rodents out of their snug burrows
underground.

POLAR AREAS

In polar areas there are practically no night animals. During
the continuous daylight of summer the whole community is more
or less active during all of the 24 hours of a "day." In winter, only
the warm-blooded animals are active at all, and these only during
a short part of each day.

EQUATORIAL REGIONS

As we go from polar regions towards the equator the summer
day becomes shorter, and the summer night increasingly longer—
so that the night animals increase in number and importance, as
we saw for the temperate forest and the desert. In the tropical forests of equatorial regions the day and night are approximately equal

in length the year around. And the twelve-hour nights are also
warm nights at all seasons. So it is not surprising that in the tropics we find the night fauna much more highly developed than in the
temperate forest, where the warm season coincides with the shorter nights.

Toad awake and active at night. Panama. (Photo by R. Buchsbaum)

Same toad asleep in its burrow during the day.

It must be added that there are many exceptions to the generalization that animals are adapted to either daytime or nighttime
conditions. First, we should recall the many animals, already described, which are primarily either diurnal or nocturnal but which
can feed either by day or by night as required by weather or by
changing needs. Some mammals, especially the big cats, like
American mountain lions or Indian tigers, are said to shift their
prowling more towards the night hours wherever they come into
frequent contact with diurnal man—a formidable enemy.

There also are animals which respond not to day and night but
primarily to dryness and wetness. Such are certain slugs and
snails. Earthworms also come out only at night and on rainy or
cloudy days. They are very sensitive to the drying effects of the
sun, but also to the ultra-violet rays of direct sunlight. An hour's
exposure to direct sunlight causes complete paralysis in some
earthworms; several hours of exposure cause death.

DAY AND NIGHT IN THE OCEAN

In the ocean there are important vertical migrations of animals
from day to night. The great masses of copepods and other animal

plankton come to the surface at night to feed on the plant plankton, then descend to greater depths (sometimes as much as 300 feet) in the daytime. They are followed in their vertical migrations by the animals that feed upon them. In the deeper layers too, there are vertical migrations of animals.

SEASONAL CHANGES IN THE COMMUNITY

The round of the seasons, in response to the yearly circling of the earth about the sun, is so much a part of our lives in temperate regions that we hardly need to have it called to our attention. Our familiar north temperate seasons are reversed in the southern hemisphere, and in both halves of the globe are limited to particular latitudes and altitudes (discussed in Chapter 8).

In polar regions we find the greatest extremes of seasonal variation and the greatest disproportion between warm and cold seasons.

In Arctic regions there are only two real seasons: winter lasts roughly nine months, and what we call spring may be the coldest months of the year, March and April taking the greatest toll of animal life. Even during the three summer months the subsoil remains frozen and cannot be penetrated by the deep roots of forest trees. But the surface soil of the treeless Arctic tundra thaws to a depth that permits the growth of lichens and mosses, small herbaceous seed plants, and in sheltered spots several kinds of low shrubs and dwarfed trees. Warm weather, with daytime temperatures of $50^{\circ}F.$, 70°, or even higher, comes with a rush. Plants almost literally burst from the ground, and the many kinds, all flowering in a brief period, produce a luxuriance that startles newcomers to the tundra. In the continuous daylight plants mature and develop seed very rapidly, so that a great variety of insects can be supported during the short summer season. Mosquitoes and flies find in the soggy soil and the myriads of standing puddles, abundant breeding places; and as they have short life cycles they are the predominant insects of the tundra, causing great discomfort to some of the mammals, among them man! Great flocks of waterfowl arrive at their nesting grounds from their winter homes in temperate regions. Innumerable small herbivorous rodents, (ground squirrels and lemmings) riddle the tundra with their burrows and at all hours of the long day pop in and out to feed. They store up the layer of fat that all permanent residents of the tundra must have to carry them through the long winter. The food stored by all these small key-industry herbivores soon goes to help fatten the carnivorous birds and mammals that feed on them: the red fox, the arctic fox, the hawks, and especially the snowy owl.

With the coming of winter the flocks of waterfowl leave. Even the snow bunting, the only song bird of the tundra, goes south. The ptarmigan loses its brown coloring and turns white (a color change common to arctic animals; see Chapter 8). It stays on as the only land bird resident in the northern Canadian tundra in winter, though it does move southward somewhat or to protected spots. It feeds on buds and seeds of the dwarf trees that protrude from the snow, and it falls prey to the snowy owl and other carnivores. The large herds of barren-ground caribou (New World reindeer) that earlier migrated northward as summer approached, and had their calves on the northern edge of the tundra, now make their way back across the "barrens," eating lichens as they go, on their way to the forests to the south. At the end of October, they stop just north of the tree line to mate. Then they move south into the protection of the forests, where they spend the winter. In spring they will start north again across the tundra. The musk ox herds are better equipped for the winter weather of the tundra than are the caribou. They can forage under the snow as well, and they have heavier fur and a more rugged, compact body. Except where men have made them extinct, they are able to hold their own on the tundra in winter.

Hibernation is seldom feasible in polar regions. Male polar bears remain active on the ice; pregnant females may retire to snow caves. Rodents cannot dig below the frost line. They prepare for winter by storing food in their burrows, and they continue foraging on lichens by tunneling under the snow. The carnivores catch rodents and use up their own body fat to fill their food deficit. It is said that the arctic fox prepares for winter by storing surplus ptarmigans in icy crevices, as we store chickens in a deepfreeze.

WINTER IN A TEMPERATE FOREST

Winter in an oak-hickory forest is not so severe as on the Arctic tundra, of course, and structural adjustments do not have to be as extreme as those of a musk ox or a snowy owl. The tempering effect of the forest foliage is gone, but the fallen leaves form a good cover for the many animals that lie dormant under the leafy

floor litter or in the soil. The defoliated trees cut the force of the
strong winds, so that the bare forest offers a moderated climate
compared with the open areas outside the forest. The white-tailed
deer herds browse on the twigs and bark of trees, and if there are
too many deer because of over-protection by man, hundreds may
starve during the winter. The red fox, finding less food in the open
woodland, prowls more on the forest floor for the squirrels that
were out of its reach in summer but now have to spend much time
on the floor searching out the acorns they cached there in the fall.
Woodchucks and bears store body fat and then hibernate (see also
page 30). During hibernation all external activity is stopped; the
body temperature drops; metabolic needs are reduced to a minimum, which can be supplied by burning body fat. Chipmunks have
little extra fat when they go into hibernation. They get up at intervals to feed, and their tracks can be seen in the snow all winter,
as can those of the rabbit, the skunk, and the raccoon, which come
out on mild days. Under the protection of the floor litter, the little
short-tailed shrew tunnels endlessly, devouring almost any sort of
edible morsel to satisfy its high metabolic needs.

All the cold-blooded forms depend upon external warmth to keep
their body machinery functioning and are now unable to remain
active. The frogs lie dormant in mud, the toads and snakes and
snails under logs or leaf litter. The insects are dead or dormant
but have provided for the new generation by leaving their young
stages secreted in rotting wood, under litter, and in crevices.
Woodpeckers, nuthatches, or chickadees, that are adept at pulling
out insect eggs or larvas from crevices of the bark of tree trunks,
stay on as winter residents. Most other birds migrate southward,
for warm blood is not enough to keep going in winter in a forest
devoid of adequate plant food or of crawling and flying insects. The
bats, which depend on insects on the wing, either hibernate in caves
or fly southward.

SEASONAL CHANGE IN DESERT AND TROPICS

As we move from polar regions towards the equator, temperature differences between winter and summer grow less, and the
important seasonal changes in the community are commonly responses to alternations of dryness and wetness. The same is really
true of arctic and temperate communities, in spite of their having
ample precipitation in both seasons. For when the environmental
temperature falls below freezing and the ground freezes hard, plan

like oak trees or hickory trees cannot obtain enough water from
the soil to offset that lost by evaporation from their broad leaf surfaces. In spite of the low temperature, evaporation continues,
especially in strong winds. Broad-leaved trees of temperate forests drop all their leaves every fall and this habit (for which we
call them deciduous trees) helps to reduce activity to a minimum
and to conserve moisture during the season when they are cut off
from their supply of ground water. Under the same physical conditions, pine trees in temperate regions are able to keep their
needle-shaped, "waterproofed" leaves, which evaporate little.

On the arctic tundra lichens remain fresh and edible under the
protecting moisture of winter snows. And the same structural
adaptations that make them independent of ground water in arctic
winters, also make them the only kind of plant able to live under
the dry conditions of bare rock surfaces in the hot summer sun of
warm climates.

Winter, in all climates, is the season in which plants cannot
thrive for lack of moisture. In our own Southwest the summer season has little if any rain. Winter is the rainy season, and the time
when plants grow lush and support an abundant animal community.
In the desert lowlands the Buckeye tree puts on its full foliage in
February, loses all its leaves in July. In the desert, as in the
arctic, only trees with needle-like leaves or other reductions of
leaf surface can remain green throughout the year. The same is
true of non-woody (herbaceous) perennial evergreen plants of the
desert, like the cacti. They have structural adaptations for re-
ducing evaporation or for storing water or for both (see also page
37). The many small annual flowering plants of the desert (like
those of the arctic) lie dormant through most of the year. Then,
following the winter rains, the desert suddenly "comes to life" in
early spring with a vast carpet stretching as far as the eye can
see, of colorful flowering annuals. The plants germinate, flower,
and seed all in a brief period of six to eight weeks. At the same
time the animal life revives. Insects multiply, toads come out of
their hot-dry-weather dormancy (aestivation), and their predators
again find their food abundant.

In the forests of the equatorial tropics the temperature is remarkably constant the year around, seldom going below 70°F. or
above 85°F. The mean temperature for January is only about 5°F.
lower than that of July. Both the dry season and the wet season

supply adequate moisture for a year-around luxuriant vegetation
and abundant animal life. Tropical rain forests are characterized
by a tremendous variety of broad-leaved but evergreen trees, and
different ones drop their leaves briefly or flower and fruit at different times of the year. So there is no season when all the trees
are bare or when none are flowering or fruiting. Specialized animals that feed only on nectar or only on fruits can find food at all
times of the year and do not migrate. Nevertheless, animals come
and go. Many birds that are winter residents of tropical forests
are summer residents of our temperate forests.

For lack of space little has been said of seasonal changes in
any but the feeding habits of animals, and, of course, feeding is
basic to the maintenance of all other animal activities. Seasonal
changes in reproduction are familiar to us in temperate regions
as the spring mating and nesting of birds. All our domestic herds
of cattle, horses, sheep, etc., are natives of temperate grasslands,
and have their young in the spring just in time to rear them during
the most favorable summer season. Arctic animals have even more
strongly marked seasonal reproductive habits. In the tropics, where
seasons are minimized, no one season is the main reproductive
season. Different species breed at different times of year. Many
species that are permanent residents of the tropics breed around
the seasonal clock. Man is a tropical mammal, and his mating
habits are as non-seasonal as those of his tropical primate relatives.

A question that comes to mind is why many birds, resident in
the tropics in winter, should come north to breed and nest in a less
friendly climate. The answer appears to be that birds which feed
only in daylight find feeding best in the long days of the temperate
spring and so can rear their helpless young in fewer days, cutting
down the length of the period when birds are extremely vulnerable.
Even more important, these migrating birds come with the spring
to occupy unexploited niches with less pressure of competition than
there was in the tropics.

Many of the seasonal rhythms in reproduction and in migration
which were once thought to be caused by seasonal changes in temperature, have been shown to be initiated by the much more regular changes in length of day. Experimental ecologists have shown,
under controlled conditions of light and temperature, that light
entering the eyes of a bird or a mammal can act as a stimulus to

the "master gland" of the body, the hypophysis or pituitary gland, to secrete hormones involved in enlargement of sex organs, changes in plumage or fur, and migratory behavior. The shortening days of early fall have been shown to be a factor in sending many birds southward, and the lengthening days of spring trigger the internal mechanisms that send some of them northward again. However, the physical or biotic factors that control physiological rhythms in animals still require a great deal of study. They are different in different animals and for each species must be determined separately. Also, the mechanism that sets off the migration in a particular species in fall is not necessarily the same as the one that reverses it in spring.

SEASONAL CHANGES IN THE OCEANS

The single most striking fact about seasonal change in the oceans is that at least 95% of the total marine environment undergoes no change that significantly affects the life of the seas. Tropical seas, especially near the equator, receive a relatively constant amount of solar energy. Polar seas do not receive enough heat from the slanting rays of the summer sun to raise by more than about $8^{\circ}F$. even the surface waters of such large masses of water. The only changes worth considering here are those of coastal temperate waters, where the surface temperature may change as much as $40^{\circ}F$. from summer to winter. Any difference at the surface decreases with depth. Off our New England coast a seasonal difference of $27^{\circ}F$. at the surface can disappear by a depth of 400 feet.

The constancy of ocean temperatures acts as a huge thermostatic device that tends to stabilize air temperatures above it, taking up heat in summer when the water is cooler than the air, and giving up heat in the winter when the water is warmer than the air above it. The moderating influence of the oceans extends to the land masses, so that land areas closest to the ocean have, in general, a much less extreme climate than those farther inland. The lowest winter temperatures and the highest summer temperatures occur in the centers of continents, as in our northwestern plains states or in Tibet.

Heat absorbed by the surface layers of water is neither radiated nor conducted from one layer to another to any significant degree. Some of the heat does reach lower layers, but mostly by actual circulation of heated surface water. It is only in shallow

coastal waters that the cold of the temperate winter and the heat
of the temperate summer are transmitted to the bottom-living
organisms.

Most plants and animals of the open seas and deeper waters
are adapted to life within a very narrow temperature range, and
when it is exceeded they die on a massive scale. Surface organisms, and those of shallow waters, are able to tolerate the marked seasonal changes of temperate waters. Hardiest of all are the
plants and animals that live between tidemarks on ocean shores.
They survive exposure, twice each day, and within a few minutes,
to alternating air and sea temperatures of both summer and winter.

Warm-bloodedness originated on the land, where the extremes
of environmental temperature in summer and winter make internal
temperature-regulation of the animal body an adaptation of great
selective advantage. A few warm-blooded animals like the whales
and seals have invaded the oceans, but in the constant temperatures of the ocean the cold-blooded giant squid can do battle in
mid-winter with its warm-blooded whale adversary. In contrast
to the land, there is no place in the ocean either too hot or too
cold for some kind of poikilothermal fish or invertebrate to be
active at all seasons.

LUNAR RHYTHMS

Many marine animals, especially polychete annelids of coastal
waters, show breeding activities that occur in relation to the lunar
cycle. Some show one or more peaks of spawning behavior that
correspond with certain phases of the lunar cycle during each of
the summer months. Others have a single annual breeding time
which is related to a particular phase of the moon. The relationships seem to be to the moonlight itself, but it is not necessarily
a simple relationship to length of exposure. In one case the swarm-
ing seems to be related to the rate at which the duration of moonlight is increasing or decreasing during the lunar cycle. Such lunar rhythms will take more study before we identify the factors involved. The best understood lunar rhythm is that of the ocean tides.

The daily tidal rhythms are twice-daily variations in water level
at ocean shores caused by the gravitational forces exerted on the
water by the moon and by the sun. Longer tidal rhythms cause more
extreme tides at times in the lunar cycle when sun and moon are

pulling together, less extreme tides when sun and moon are pulling in opposition to each other. All these fluctuations in water level are critical factors in the life of animals of tidal shores.

When the tide is out animals of the shore are subject to extremes of land temperature in summer and in winter, to a sudden decrease in salinity should it be raining, to extreme drying during the hot days of summer, to a cessation in feeding activity in most forms, to mechanical damage when the tide rushes in again, to the predatory activities of shore birds, and to many other hazards from which the watery medium protects them. The range of physical conditions that occur on tidal shores, all within a few yards of each other, would on land usually be separated by hundreds of miles. (See also the discussion of zonation on rocky ocean shores in Chapter 8.)

Chapter 6

ECOLOGICAL SUCCESSION

THE CHANGES that occur in a community from day to night
and from summer to winter are cyclic changes, recurring over
and over again in a constant pattern. There are other kinds of
community change that follow a definite pattern but which do not
recur with periodic regularity. Such are the long term changes
that follow the changing climates of the various geological ages as
the land masses are uplifted and then subside, as new mountain
ranges are formed, and as ice ages come and go. There are also
the gradual changes in the organisms themselves in the process
of organic evolution. So the communities we observe and describe
today, with their particular composition of plant and animal species, are not the communities of yesterday, geologically speaking.
Students of paleontology and paleogeography have learned a great
deal about the kinds of organisms that there were in the past and
and about their distribution over the earth. We cannot consider
them here, but they are well described in many very readable
sources; some are listed in the Bibliography at the end of this book.

The kind of community change we shall consider here takes
place without any large scale change in climate. It is initiated
primarily by changes in the surface configuration or the physical
condition of the soil, which we call physiographic changes. These
may be the erosion of a whole landscape, the filling up of a lake,
the deepening or silting up of a river bed, the piling up of sand at
the edge of a lake. Or the changes may result from a local catastrophe such as an earthquake, a forest fire, or the draining of a
swamp. Whatever the original physical causes, what finally takes
place is also dependent upon the activities of the plants and animals
themselves.

Given any particular set of physiographic factors, combined with
a certain climate, the sequence of events that follows has a predictable pattern which we call ecological succession. Each of the
stages in a succession is an integrated community with the community structure described earlier and with one or more dominant

organisms.

Some successions occur on a very small scale, as with the protozoans of a temporary rain puddle or a water-filled tree hole.
They may be over in a matter of days, or a few weeks or months.
A rotting log may go through all its successive stages in a few
years, and we can follow the whole process in a single log. A
large-scale succession, however, involving a change from bare
ground to mature forest, may take thousands of years. In such a
case we cannot expect to witness the changes that take place in any
particular spot. We can only infer these changes by examining
what are apparently successive stages in different parts of the
changing area. We shall see examples of this later in the descrip-
tions of the stages that intervene in the development from sandy
beach to beech and sugar-maple forest.

A tree hole holding a little water quickly gains decomposing organic matter and
a variety of animals. The community undergoes ecological succession and may
support many other animals, depending upon its size, duration, and kinds of organic matter. This one in Panama contained half a pint of water and bacteria,
flagellates, ciliates, amebas, rotifers, gastrotrichs, nematode worms, mosquito
larvas, and tadpoles. (Photo by R. Buchsbaum.)

The first stage in a succession is a <u>pioneering stage</u> in which the first organisms must be hardy plants that can grow on bare soil or bare rock, or if in water, on inorganic salts. When the plants have established themselves and so provided a food base, small herbivores may invade the area, but they are usually of a kind that can seek shelter under stones or can burrow in the bare ground. After a time the pioneers have so changed the original conditions that they are no longer adapted to live in the area. They are gradually pushed out by newcomers better adapted to the now changed conditions. The new assemblage of organisms lasts only so long as its members can tolerate the constantly changing conditions they produce by their own activities. Then they in turn are replaced by a new set of organisms. This process of self-induced change continues until at last there is produced a community which is able to propagate itself indefinitely because it is in dynamic equilibrium with both soil and climate and because it is the fullest expression of plant and animal life that can be supported in the area. Such a stable community is called the <u>climax community</u>, and it persists because it does not create conditions which are unfavorable to itself or which favor the invasion of better adapted forms.

The whole series of communities that lead to any one climax is called a <u>sere</u>; each stage in the sere is a <u>seral stage.</u> Several examples of successional seres will be given to bring out the principles.

PROTOZOAN SUCCESSION

The protozoan successions that occur in natural situations like temporary pools and tree holes, or in artifacts like cemetery urns, can be simulated in the laboratory. There they lend themselves to convenient examination and to experimental modification.

A small amount of ordinary hay is boiled in a pan of water. The resulting brownish "hay tea" is then filtered off and allowed to cool in the laboratory. At first the "tea" is practically sterile and contains only undecomposed organic matter derived from the hay. At this point we inoculate the culture with a small amount of natural pond water with its contained organisms. At the same time the spores of bacteria settle out of the air into the solution—a totally unexploited nutritive medium. The clear fluid becomes cloudy as the bacteria multiply. During this period of almost uncontrolled multiplication, the <u>bacteria</u> carry on fermentation, with a consequent production of acid. The acidity slowly rises, and the bacteria convert the tea (which at first could not support anything but bacteria) into a form of food which is suitable for protozoan metabolism. There follows a rich growth of protozoans whose order of peak multiplication is affected, not by the number of protozoan individuals of any one kind that may have been inoculated into the medium, but by the changing environmental conditions.

First to multiply rapidly are the primitive flagellates. They feed not only upon the substances which the bacteria release into the water, but also upon the bacteria themselves. The flagellates make such inroads upon the bacterial population that they literally "eat themselves out of house and home" and almost disappear. The decline of these flagellates is also speeded by the rise of a ciliated protozoan related to Paramecium, called Colpoda. Whether the high acidity at this time favors the multiplication of this form over others, whose members may be present in even greater numbers, is not known. In any case, decreasing acidity of the medium is paralled by the sudden decline of Colpoda and by the rise of other ciliated protozoans, the hypotrichs (so-called because the cilia are larger on the under side). As the hypotrichs lose their ascendancy, Paramecium attains prominence, and lives on in fairly good numbers for months. Along with Paramecium we can usually watch the rise of Vorticella, a stalked and temporarily-fixed protozoan whose cilia are strongly developed around the mouth. Vorticellas disappear from the surface of the culture shortly after the last Paramecium retires to a life of obscurity at the bottom. Amebas appear very early or not until after several months.

This succession may die out soon afterwards unless we intervene to replenish the food supply. The new food will set the cycle going again—in the same sequence as just described. However, if the culture is left undisturbed a kind of permanency is usually attained. Green algae finally multiply and provide a constant source of energy, making life fairly stable. All of the animals mentioned above persist in small numbers among the algae. And in addition, the many-celled rotifers and crustaceans make their appearance. These do not multiply during the initial or pioneer stages but usually during the more mature stages of a culture. As long as the algae receive enough light to form a food basis for this microcosm, the numbers of animals of different types will remain about the same for many months or even years. This relatively stable condition is the climax stage.

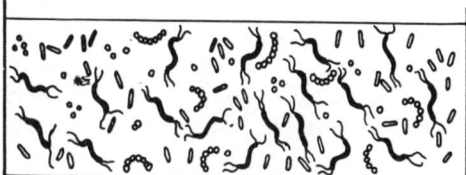

Bacteria enter from the air (or from a seeding of pond water) and multiply rapidly. Other plants and animals enter also, but do not develop immediately. This is the **pioneer stage.**

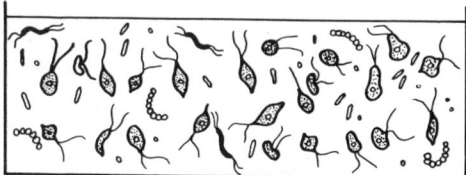

Flagellates are the first protozoans to appear. Many bacteria remain.

The succession does not follow a typical pattern unless the seeding contained the appropriate organisms.

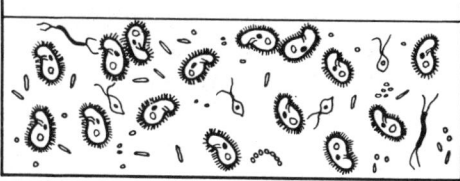

Colpoda rise as the flagellates decline. Some bacteria and some flagellates remain.

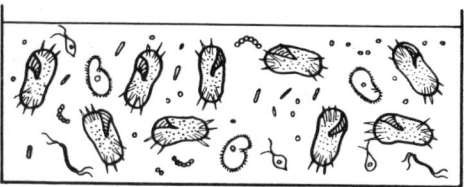

Hypotrichs now grow in the medium which becomes unfit for maximum growth of Colpoda, but a few Colpoda remain.

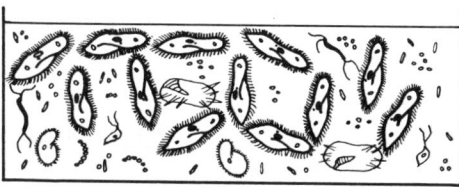

Paramecium attains prominence as the hypotrichs decrease in numbers.

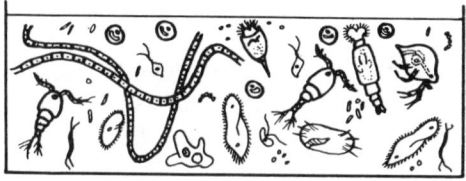

The **climax** stage is a balanced microcosm: green algae, rotifers, crustaceans, and amebas are recent, but a few representatives of each preceding stage remain.

Protozoan succession may be observed in a jar of water containing a source of organic material and a seeding of pond water organisms.

While the sequence of events in this simplest of the well-known
ecological successions can be seen by anyone capable of using a
microscope, the interpretation is still very imperfectly worked
out. There is no doubt that the kind and amount of food available
is an important factor. Changes in acidity probably play a role.
Some definite evidence has been obtained that the excretions of the
protozoans themselves are significant. If Paramecium is grown in
pure culture (no other animals being present) in water containing
large amounts of the excreta of former generations of Paramecium,
the animals will reproduce more slowly than they normally do,
even though the food supply is fresh and quite adequate. But Paramecium reproduces at a normal rate in water contaminated with
the excreta of hypotrichs. This suggests how Paramecium can rise
to a maximum in the wake of a declining growth of hypotrichs and
then decrease in number as its own waste products accumulate in
the culture. When animals exploit an environment, their own life-
activities tend to make the surroundings unsuited for their continued multiplication. However, the very changes which render the
environment unfavorable for their own survival pave the way for
some other form of life.

FALLEN TREE SUCCESSION

Much more complex than the protozoan succession, and usually
lasting over a period of years, is the succession of animal communities inhabiting a fallen tree. A newly fallen beech log is attacked by the pioneering boring beetles which make tunnels in the
solid wood. Through such openings there enter other small animals,
fungus spores, and water. The tunneling and feeding of the beetles,
the growth of the fungi, and the freezing of the water in winter all
help to loosen the bark and to soften the wood. Under the loose
bark we can then find many small snails, a slug or two, the larvas
of fungus-gnats, and especially the flattened types of beetle larvas.
As the wood is softened by the pioneering animals it no longer supports boring beetles, and they leave, to be replaced by fungus-eat-
ing beetles and by the larvas of click-beetles that can burrow only
into softened wood. Finally, the bark is completely gone, and the
forms that sheltered there no longer remain. Molds and bacteria
reduce the pulpy wood to a mound of soft humus in which we can
find ants, earthworms, ground-beetles, millipedes, and other animals that live also on the forest floor.

Log succession is illustrated in first and last stages in this Vermont forest. *Above,* a newly cut and discarded birch log shows no obvious signs of decay. *Below,* the remains of a thoroughly rotten fallen tree are beginning to blend into the humus of the forest floor. (Photos by Ralph Buchsbaum.)

SUCCESSION FOLLOWING VOLCANIC ACTIVITY

Ecological succession on a much grander scale occurred after the eruption in 1883 of the volcano on the island of Krakatoa in the East Indies. In a series of terrific explosions, half the island was blown away. Some 36,000 people died in the great tidal wave that was sent surging onto neighboring islands. Nine months after the eruption, a French botanist could find no trace of plant or animal life except a single spider brought by the wind. Two years after this visit, a party led by a Dutch botanist found on the beach of Krakatoa many of the plants that grow on tropical seashores. Inland from the beach there were only the pioneering blue-green algae and bacteria of the soil, some scattered mosses and many ferns—all organisms spread by spores that could have been brought by the wind from the neighboring islands of Java and Sumatra. Grasses were among the few seed plants. Thirteen years after the explosion the island looked green again. The shores, fringed with cocoanut trees, had a better developed vegetation than did the interior, but there were scattered trees inland, wild sugar cane, and four species of ground orchids. By 1906 the island was densely covered with vegetation, most of it grass. The cocoanut trees on the shores were denser, and figs were established; but trees were still scattered. Animals, too, had arrived, almost all flying forms: ants, mosquitoes, birds, and bats. Lizards were present and could have arrived on floating wood. In 1920 trees occupied about half the surface of the island. And by 1930 the whole island was again a dense forest, though a young one. From bare soil to forest in a little less than 50 years is a remarkable change, considering that all the plants had to arrive on the island by natural means. Only in the tropics could such a rapid natural reforestation take place.

Ecological successions started by volcanic activity are fairly common. At certain intervals in the past great outpourings of lava have come from extensive fissures in the earth. Lava flow following lava flow have covered large parts of Oregon, Washington, and Idaho (about 200,000 square miles) to a depth of about 2000 feet. The climate in the northwestern United States is not tropical like that of Krakatoa, and under the less favorable physical conditions succession has proceeded at a very much slower rate. In the Cascade Mountains of Oregon, at an elevation of about 10,000 feet above sea level, extensive areas of black lava still appear entirely bare, as if the flow had but recently occurred. Farther down on the eastern slopes, where the climate is milder, pioneering lichens,

Lava supports a sparse vegetative cover in almost any climate, but what it does support depends upon the amount of moisture available. *Above,* coniferous forest on lava beds in the Cascade Range near Bend, Oregon. *Below,* cactus and sage on lava beds in New Mexico. (Photos by Ralph Buchsbaum.)

plus weathering of the rock have produced sufficient soil to provide for the return of pine or juniper forests. In the low, warm,
protected valleys of eastern Washington, the lichens were long
ago replaced by sagebrush and numerous grasses. In southwestern
New Mexico, under very dry conditions, ancient lava flows are
now sparsely covered with cactus and yucca. So we see that the
same kind of physiographic change, occurring in different climatic
areas, produces different sets of vegetation.

POND SUCCESSION

From what we already know of the protozoan succession, it is
not difficult to picture the similar successions that are started
wherever pools of water containing organic matter are formed following a rain: in the crevices of rocks, in the cup-like cavities at
the bases of large leaves, in discarded bottles or tin cans, or in
mud puddles. These scarcely ever reach a climax stage, being
brought to an abrupt end in a few hours, a few days, or a few weeks
by the evaporation or drainage of the water. This fate does not be-
fall larger bodies of water. When bodies of fresh water are too
small to be called lakes, we call them ponds. When ponds are
occupied by vegetation to the point where there are no longer large
areas of open water, we call them marshes. Lakes, ponds, and
marshes are relative terms, and no sharp distinctions can be made.

Lakes, ponds, and marshes are only stages in a continuous
process. The evolution of a body of water from lake to marsh
takes thousands or even hundreds of thousands of years, depending
upon the size of the lake. We cannot observe the process directly,
and therefore we employ a method, so commonly used in science,
of finding most of the links or stages and putting them together to
give a complete story.

The predecessor of Lake Michigan, for example, stood at a
level about 60 feet above the present lake. As the lake receded,
wave action cut away cliffs and formed beaches of sand and gravel
at various levels. The water level fell lower and lower and left
beaches as long ridges on the exposed lake plain at the southern
end of the lake. And in the depressions between the ridges were
formed a series of 95 parallel, narrow, shallow ponds, usually
several miles in length, the older ponds being farthest from the
present shore of the lake and the youngest being very close to the
present shore of the lake. The ponds lie within the city limits of

Gary, Indiana, and may be seen from the road leading to the Indiana Dunes from Chicago. In recent years many of them have been drained and no longer exist. But they were once very intensively studied by Shelford and many others, and from these studies on a large number of different ponds of various ages has been derived a detailed story of the evolution of a single pond. Only a bare outline of this story will be given here to illustrate the general way in which ponds in other places may undergo succession.

When first formed, such a pond has a sandy bottom, and we say that it is in its pioneer or bare bottom stage. The very earliest phases of this stage are difficult to find in nature. Presumably, bacteria, protozoans, microscopic algae, small crustaceans, and rotifers first form a plankton community. After a time the floating plankton grows rich enough to support larger forms which have entered while the pond was still connected with the lake. The fish (black bass, bluegill, and speckled bullhead) all make nests on the sandy bottom; the male fish guard the nests. Snails scour the bottom for green algae. Caddisfly larvas crawl over the bottom, feeding on microorganisms, and use sand grains to build the peculiar cases in which they live. Mussels strain plankton from the water, but they require a sandy bottom for their mode of locomotion; they cannot maintain themselves in an upright position if vegetation covers or binds the sand. All of the animals of this stage are not just tolerant of the bare bottom; they are dependent upon it.

After a number of years the decayed bodies of plants and animals have formed a layer of humus over the bottom, enabling the large and branching green alga, Chara, to grow. The Chara covers the bottom and contributes abundant humus, making life impossible for the bare bottom types. As the Chara never reaches the surface of the water, we speak of this new community of animals and plants as the submerged vegetation stage. Chara is a siliceous (glassy) alga and is probably not eaten by pond animals. It serves as a resting-place or as a hiding-place. Dragonfly and mayfly nymphs burrow in the muddy bottom, but aquatic insects are scarce. Many small crustaceans and a few crayfish are found. The nest-building fishes are gradually replaced by others (golden shiner and mud minnow), which lay their eggs on the Chara. The snails are replaced by a species which finds its food on the Chara. The caddisfly larvas of the preceding stage have been replaced by another species which creeps over the Chara and makes its case not only of sand grains (as did its predecessor) but also of pieces of humus.

A different species of mussel occurs. The submerged vegetation community is characterized by forms which are independent of both surface and bottom, but are dependent upon bottom vegetation for nesting places and for shelter.

The Chara continues to fill in the pond at the rate of about one inch a year, providing a layer of soil for plants having greater demands than the pioneering Chara. Cattails and bulrushes invade the edges of the pond and water-lilies may grow nearer the middle. These plants have their roots in the humus-covered bottom, but reach above the surface of the water, affording a habitat for semiaquatic forms which must climb to the surface to obtain oxygen.

Thus we have a new community developing which is adapted to the emerging vegetation stage. Those few persistent mussels that lingered on while the Chara was still sparse are entirely gone now; their niche is filled by smaller bivalved mollusks which can live in the humus of the bottom or can climb about on the vegetation. The burrowing dragonfly numphs are replaced by species that climb on the submerged parts of bulrushes and on submerged twigs that have fallen in from shrubs overhanging the pond's edge. The gill-breathing snails are replaced by lung-breathers which climb to the surface to renew their supply of oxygen. Diving spiders rest on the exposed bulrush and cattail stalks and dive after the abundant mayfly and dragonfly nymphs. These nymphs pass their early stages clinging to the submerged vegetation and then climb to the surface when they are about to shed their nymphal coverings and fly off. As we have now come to expect, the caddisfly larvas are represented by a species well adapted to the changed conditions; the case is made of grass blades or other fragments of water plants. Aquatic larvas are abundant and supply food for a large population of diving beetles and numerous other adult carnivorous insects. These are an intermediate link in a food chain that ends with larger forms like crayfishes, frogs, salamanders, turtles, and a few fishes such as bullheads. Annelid worms live in the mud on the bottom, extracting organic material from the humus. Leeches prey on their fellow-members of the pond, but take occasional meals from ecologists who wade in to study succession. The emerging vegetation stage differs from its predecessor chiefly in harboring many animals that require a dry support at some time in their lives. Also, the more stagnant, vegetation-choked pond has a decreased supply of oxygen, and forms which require a high oxygen content cannot survive. Crayfishes solve this difficulty for their

young by carrying the developing eggs around with them instead of
depositing them in the mud at the bottom where there is little oxygen.

After a time the vegetation which made life possible for the animals of this stage, together with the waste products and dead
bodies of the animals themselves, so fill the pond that there are
no longer any large-sized areas of open water. The senescent pond,
occupied by grasses and sedges, becomes a marsh. (If occupied by
trees, like the tamarack or cypress, we call it a swamp. Many
large marshes or swamps are fed throughout the year by springs
or by some large body of water and have relative permancy. The
one we describe here has no water supply except local rainfall.)
When well-covered with vegetation it becomes so muddy and shallow, that the truly aquatic animals die. Frogs, salamanders, crayfish, and leeches remain, but the fish cannot. Turtles bask on the
logs that protrude above the surface. Garter snakes glide about at
the edges, feeding on frogs or on dead or dying fish.

When filling of the pond at last brings the bottom above ground-
water level, and the pond dries completely during the summer, we
have the temporary pond stage. Only animals which are adapted to
withstand complete drying in the summer and freezing temperatures
in the winter can reside in the temporary pond. Bacteria, protozoa,
molds, yeasts, algae, and rotifers meet this problem by the formation of heavy-walled resistant spores and cysts. Snails with-
draw into their shells, seal themselves against loss of water by
secreting a film of mucus over the shell opening, and then lie bur-
ried in the mud, where they withstand freezing. Amphibians and
crayfishes also bury themselves in the mud. Many forms cannot
survive these conditions as adults, but their eggs, which are small
and protected by a heavy coat, survive the summer and winter
months and produce the species the following spring. The common
fairy shrimp, a tiny crustacean, lives mostly in small temporary
ponds—never, it is said, in ponds that support fish. It thus exploits
a niche unavailable to animals with eggs that cannot survive drying
and freezing, and at the same time avoids animals of a size that
can prey on fairy shrimps.

Animals and plants of the temporary pond continue to add debris
which fills the pond and hastens their eventual expulsion. After a
time the pond is wet only during March and April; during the rest
of the year it is part of a dry prairie and is occupied by land ani-

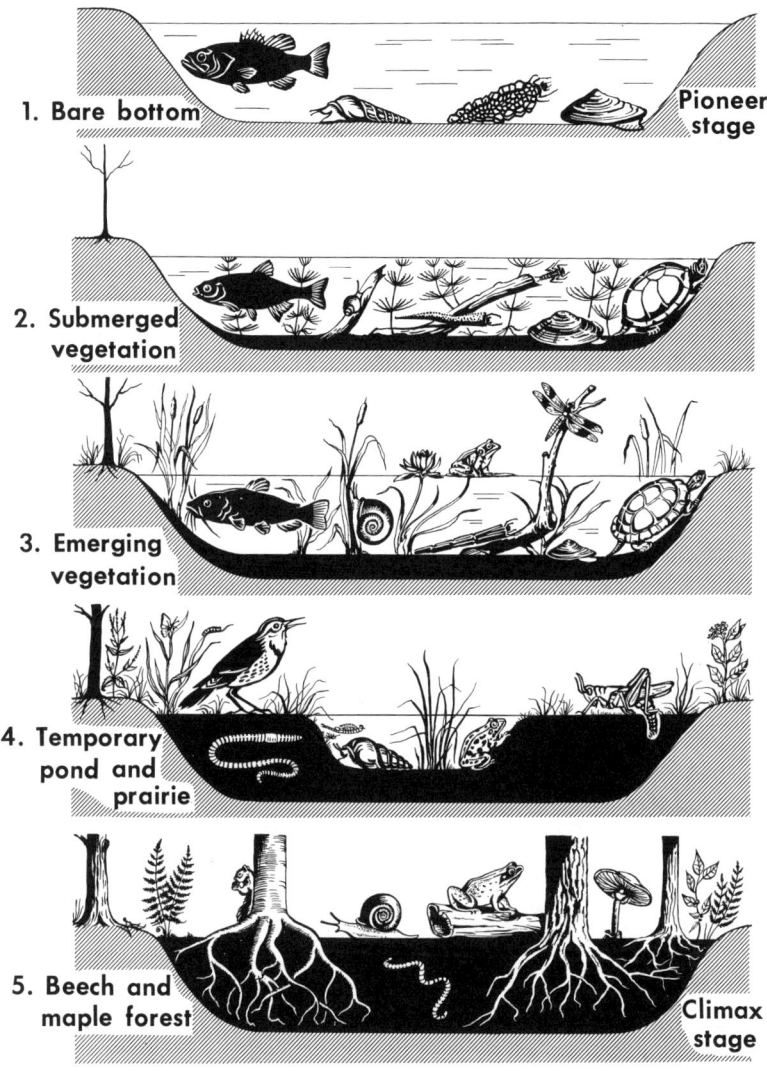

1. **Bare bottom**   Pioneer
                                                                        stage

2. **Submerged
    vegetation**

3. **Emerging
    vegetation**

4. **Temporary
    pond and
        prairie**

5. **Beech and
    maple forest**   Climax
                                                                        stage

Pond succession, simplified. The stages shown here are based upon events
which are thought to occur in the ponds at the south end of Lake Michigan.
Other ponds in the Middle West will go through much the same succession.
The succession in ponds with a very different climate will be different in detail; and, of course, the climax will be different. Humus laid down since the
beginning is shown in solid black.

mals and plants. At this time we speak of the pond as being in the
low prairie stage.

When the surface of the low prairie is built up by plant and animal debris to a level that remains at all times above ground-water,
the bulrushes go and midwestern high grass prairie develops. But
the story of a pond at the southern end of Lake Michigan may also
have another ending. This region is a transition zone (or ecotone)
between the great forestlands of the eastern part of the country
and the grasslands that lie to the west of Chicago. Depending upon
topography and soil conditions, and also upon the past history of
the area, either prairie or beech-maple forest may be the climax
stage. If the pond is in an area that naturally succeeds to forest,
the marshy stage of the pond gives way not to low prairie but to a
moist thicket of various shrubs like the buttonbush. Then come
oaks and other trees and finally the beeches and maples that stabilize the community.

A pond in the *emerging vegetation* stage in pond succession. The margins of
the pond are further along in the succession than the center, for several shrubs
may be seen here and there at the edges of the pond. As time goes on and the
pond becomes filled in with humus, it will become either a pasture (or cultivated field, depending upon the farmer who owns the field) or, if left alone for
a long enough time, a forest like that seen in the background. (Photo taken in
northern Minnesota by Ralph Buchsbaum.)

SAND DUNE SUCCESSION

   Although the pond succession was described as if we were watching the evolution in time of a single pond, in describing the evolution of a sandy beach we will stick closer to the actual method of
study and will observe the evolution of the beach in space. To do
this let us imagine that we are making a trip to a sand dunes region; it could be almost any with which we are already familiar.
Just to be specific, let us visit the dunes at the south end of Lake
Michigan, where the succession has been most studied. These
dunes are the result of the water currents in the lake, which erode
the western shore and deposit sand at the southern end of the lake.
During the winter months, high waves carry the sand far up on the
beach. During the summer this sand dries, and when the winds
are off the lake, the sand is blown inland. Wind-blown sand tends
to drop when the wind is slowed by any obstacle, be it a tree, a
blade of grass, or a pebble. Small sand hills, thus formed, stop
more and more sand and continue to grow—finally reaching a height
of 200 or 300 feet. The dunes are not stationary but tend to move
about as the wind blows sand off the windward side and deposits it
on the leeward side.

   We arrive at the water's edge on a warm sunny day, following
a summer storm. Along the wet sand margin, as far as we can see,
is a pile of drift washed up by last night's high waves but now just
beyond the reach of the waves. The drift contains a large amount
of plant debris and is alive with wriggling butterflies, potato beetles, ladybird beetles, and horseflies—all blown into the lake by
the strong winds and then cast ashore by the waves. Many of the
insects are dead, but some are slowly recovering. There are a
few fish, all of them victims of lampreys or of some misfortune
which rendered them defenseless against the landward sweep of
the strong storm waves. If we walk along the drift-line we can
count hundreds of dead birds and perhaps even an occasional mammal. But we are not the first to make this interesting find. The
flesh flies and various scavenger beetles have come before us and
are feeding on the dead animals. The flesh flies lay their eggs in
dead fish, which they are able to find with such regularity that
they use them as their breeding-ground. The spotted sandpiper
feeds on the maggots that hatch from the eggs of the flesh flies.
Tiger beetles prey on small beach insects. At night toads, snakes,
white-footed mice, and skunks come down for a midnight "snack."
The animal community of the wet sand margin is a temporary one,

composed of scavengers which come to feed, or carnivores which eat the scavengers, but which live out of the reach of the waves in the drier sand of the middle beach or beyond.

The middle beach is the wide flat strip of sand that lies between the wet sand margin and the first small dunes. The beach is constantly being widened by sand and debris deposited by the waves of severe winter storms; it thus becomes the final resting-place for driftwood which is first washed up on the wet sand. In the summer the constantly shifting, hot, dry sand provides an environment only slightly more stable and inviting than that of the wet sand. None but very adaptable animals, hardy pioneer types, can meet the rigorous requirements of life on the middle beach. These forms, as we have seen, are the predators and scavengers of the drift-line. They are not adapted, as are true desert animals, to withstand the drying action of the sun, but take advantage of the presence of driftwood, under which they find relatively cool and moist conditions. When food is scarce on the water margin and few scavengers are to be found there, we can see them all "at home" in the middle beach: various bugs and beetles hide under the smaller sticks, and toads, mice, and sandpipers under the large logs. In addition, many of the logs serve both as shelter and food for the wood-eating termites. A few ants may be found in the logs, but they are not common. The burrowing spiders and the larvas of tiger beetles solve the shelter problem by digging down into the cool moist sand that lies just below the dry surface. The activities of these burrowing forms, as well as the movements of the predators, help to scatter the humus from decaying plant and animal bodies and to mix it thoroughly with the sand.

High up on the middle beach there is enough humus mixed with the sand to support a few pioneering plants. The plant pioneers must be perennials, they must resist extreme drying, they must require a minimum of soil minerals, and they must bind the sand with their roots to prevent it from shifting and overwhelming them. Among the plants able to fulfill these requirements is the sand grass. The leaves and roots of the sand grass serve as obstructions against which the first small dunes are formed. As these dunes move farther and farther from the windswept middle beach, their speed of movement gradually slows. When the rate of movement drops below a certain point, the dunes are "captured" or immobilized by the binding roots of the sand grasses. These are the fore dunes. From season to season, the decaying roots and leaves of the sand grass add more humus to the sand, and the sea rocket begins to appear here and there in the moister spots, followed by the willow and the sand cherry, that grow from seeds brought by birds. Still farther inland, in the damp depressions between the dunes, the first cottonwood (a poplar) seedlings appear.

Up to now the succession has largely been dominated by the influence of the shifting sand. From this cottonwood stage on the movement of the sand becomes less and less important and the chief influence is that of the increasing soil humus, which provides more minerals for growing plants. As the plants increase in number they provide more moisture, more shelter, and more food for the animals which follow them. It is true that the animals contribute a large share of the organic debris which fertilizes the soil and that the activities of the digging forms are important in carrying the humus far below the surface of the sand. But without the pioneering activities of the plants in binding the sand and in contributing organic debris, the animals could never get beyond the middle beach stage. In addition to the cottonwoods, this region still harbors the sand grass, willow, and sand cherries. The deeper depressions between the dunes may become temporary ponds and support a luxuriant growth of sedges.

The leaves of the cottonwood trees afford a convenient home for the crab spider. The willows furnish food for beetles, and in early spring the willow blossoms are visited by pollen-gathering insects, which in turn attract the kingbirds. The sand cherry is attacked by aphids, which are fed upon by certain flies. The cherries attract birds. Grasshoppers, locusts, midges, mosquitoes, and flies (which breed on the beach) rest on the grasses and low shrubs. Sparrows feed on the seeds of the grasses and herbs. The spiders which often built their burrows in the drier sand of the middle beach and fore dune are still quite numerous here. The burrowing larvas of the tiger beetle are common but no longer of the same species as those of the middle beach. The characteristic representative of the fore dune and cottonwood zones is the white tiger beetle. Ants, which are relatively rare on the middle beach and fore dune begin to appear in appreciable numbers. But perhaps the most characteristic animals are the digger wasps. These wasps are somewhat gregarious in their habits, and often build their nests in groups. On sunny days they collect flies from the beach and store them in their burrows; the storing of these flies contributes to the accumulation of debris below the surface. The cotton-

wood community is characterized by swift, predatory animals.
They hunt food in the daytime and spend nights and cloudy days in
their underground burrows, where they also breed.

The invasion of the region of large cottonwood trees by scattered pine seedlings and bunchgrass forms a transition zone, known
as the old cottonwood-young pine seedling dune, which differs from
most transition zones in being a rather well-marked belt with a
fairly characteristic animal community of its own. The digger
wasps are still abundant. The tiger beetles, as we have come to
expect, are represented by a new species which we may call, for
lack of a more distinctive common name, the large tiger beetle.
This form differs from its predecessor not only in size and in
other physical characters but also in its habits. Its burrows are
not simple vertical passages but are curved near the top like a
stovepipe.

Beyond the transition zone is the dense pine forest, and from
this point on the numbers of different plants and of animals are so
great that only a few can be mentioned. We must bear in mind that
despite the lack of detail given, the animal and plant communities
are becoming increasingly dense and complex as we go farther and
farther from the lake.

When the pioneering cottonwoods have added enough humus to
the soil so that the sand is definitely darkened, the consequent increase in soil minerals and in the water-holding capacity of the
soil, plus the increased shade from the hot summer sun afford the
necessary conditions for the growth of the young pine seedlings.
The more successful pines outgrow the cottonwoods and dominate
the situation. The cottonwoods cannot grow in the shade of the
pines and gradually disappear. The forest-floor supports not only
bunch grass but also Solomon's-seal, mint, goldenrod, and a variety of other flowers, grasses, and small shrubs. A number of
burrowing forms (wasps and spiders) of the preceding stage persist
here, digging their burrows in the sand just below the pine needles
which cover the ground. The increased humus and moisture are
strikingly reflected in the ant population, which jumps from three
inconspicuous species in the cottonwoods (living mainly in non-
typical places like damp depressions and under logs) to a rather
active representation of eight species, each confined to some
special habitat in the pine forest: open sand, sand under pine
needles, and under logs, etc. With the increased shrubs the num-

ber of species of grasshoppers increases and occasional cicadas may be seen. The tiger beetles are represented by the bronze tiger beetle, the larvas of which build straight, cylindrical burrows. The six-lined lizard, the blue-racer snake, and the pond turtle all lay their eggs in the sand.

The pine trees themselves support a whole new group of animals. The twigs, the leaves, the trunk—each have their own types of leaf-cutter or wood-borer. And the dead and dying trees are attacked by bark beetles. Many birds nest in both the dead and live trees.

The abundant humus, contributed by the constantly-shed needles of the pines and by the activities of the ever-increasing number of animals, eventually permits the development of an oak forest. Under the protecting shade of the black oaks, the blueberry, the grape, the choke-cherry, and the rose come in. The animals are too numerous to be given in detail, but in general it may be said that the digging forms, characteristic of the bare-sand stages are almost gone. Their digging activities have aided in changing the soil so much that they can no longer live here and in their place we see a marked increase in number and diversification of animals which depend for their shelter on the trees and on the vegetation covering the ground. Earthworms are just beginning to come in and to take over the soil-turning activities of the digger wasps. Ants find a great variety of habitats in the black-oak forest and rise to maximum numbers and activity here. But most characteristic is the increase in tree-living animals. The cottonwood trees were well exposed to the hot summer sun and to the strong lake winds and supported some insects and only a few hardy spiders. The pines had some borers and a larger spider population. But the lower branches of the black oak forest are quite sheltered, and a vast army of leaf-cutters, wood-borers, caterpillars, gall-formers, plant lice, walking-sticks, and other insects have arrived to feed on the trees or on the animals that feed on the trees.

The falling leaves of the black oaks rapidly increase the soil humus; the dense growth of the trees increase the shade and humidity; and the black oaks are gradually displaced by the white oaks and then by red oaks and hickories. Meeting these changed conditions is the green tiger beetle. Large forest mammals are present in numbers. The groundhog and red fox nest in burrows, as did the wolf in former times. The rabbits (really hares) of the nearby prairies often winter in the forest. Wood mice nest under

fallen trees. Squirrels are numerous, and birds nest both in the
trees and on the ground. Snakes and salamanders are seen occasionally. But most characteristic is the increase of the true forms of
the forest floor: earthworms, crickets, millipedes, snails, and
slugs, living under rotting logs and in the dead moist leaves.

In our walk from the middle beach to the oak forest we have
seen an increasing vegetation which affords an increasing number
of habitats to an ever-growing variety of animals, reaching a
peak in the oak forest. When we begin to find beech and maple
seedlings we know that the oaks and hickories will soon give way
to the dense beech and maple forest beyond. The beech and maple
forest continues the tendency to increased humus (with accompanying water-holding capacity of the soil) and protection from wind,
heat, and drying in the summer. The dense growth of beech and
maple trees makes the forest-floor so shady that the rich growth
of small vegetation dies out. With it go the immense variety of
insects and mammals which depend on this underbrush for shelter,
breeding places, or food. Wasps, tiger beetles, ants, and other
forms which did not depend upon the shrubs for shelter, cannot
live here because they cannot dig under the heavy matting of dead,
water-soaked leaves that cover the forest floor. So the number of
animal niches declines in the climax community of this sere.

The beech and maple forest is characterized by typical "damp
forest" conditions. The forest floor is covered with logs and fallen
leaves but is relatively bare of shrubs or animals. The earthworms
continue, and under the leaves are scattered snails, centipedes,
millipedes, and crickets. In general, the animals are less abundant under the leaves than under the logs, where we find many
slugs, snails, centipedes, millipedes, spiders, cranefly larvas,
beetle larvas, numerous adult beetles, and ants. Each fallen log
presents a small succession of its own, as we have previously
seen. Salamanders and wood frogs are numerous.

Few new plants can grow up in the dense shade of the beeches
and maples, and since these trees can continue to reproduce themselves indefinitely, this type of community remains essentially
the same—if nothing occurs to upset it.

Local changes in the wind can no longer affect the densely over-
grown dunes which lie beneath the beech and maple forest, but at
any region from the fore dune to the oak forest a temporary change

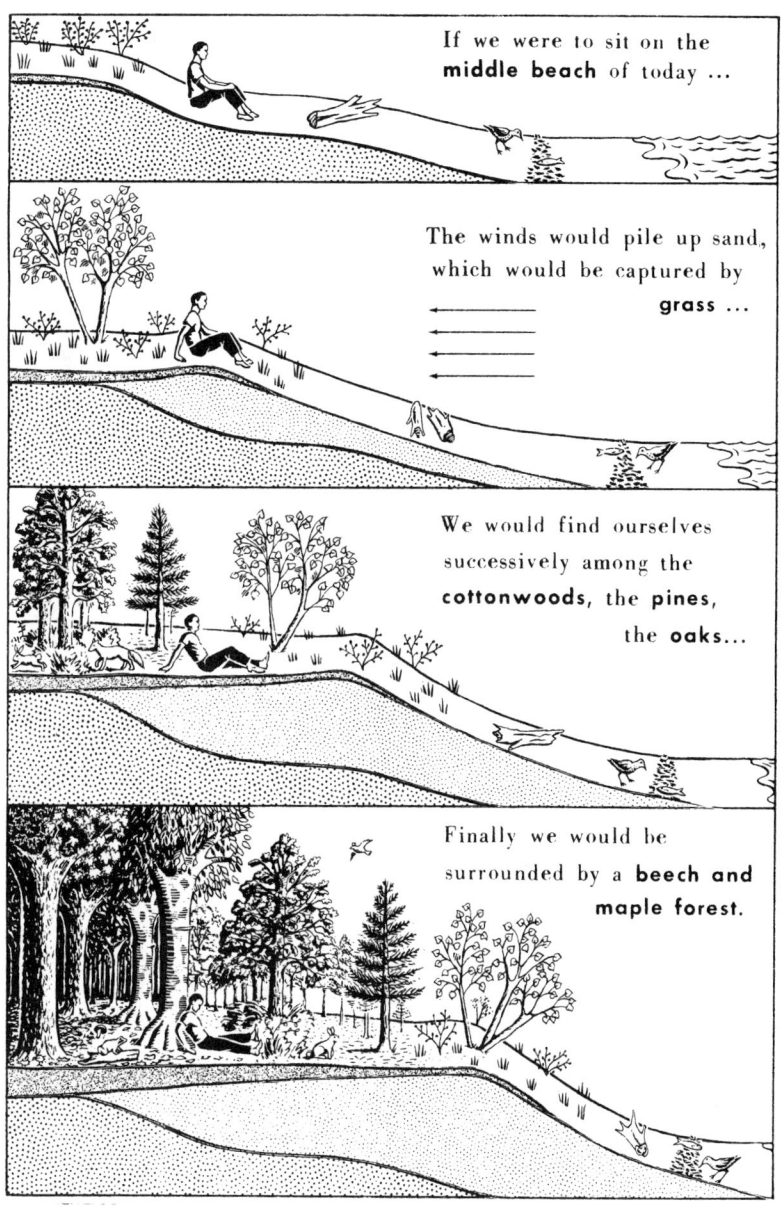

If we were to sit on the **middle beach** of today ...

The winds would pile up sand, which would be captured by **grass** ...

We would find ourselves successively among the **cottonwoods**, the **pines**, the **oaks**...

Finally we would be surrounded by a **beech and maple forest.**

Sand, present at the time we first sat on the middle beach.

Sand, washed up by the waves and blown by the wind since we first sat on the beach.

Humus, added by plants and animals.

Dune succession (based on succession at the south end of Lake Michigan).

Dune succession may be seen almost any place where there are sand dunes. Here at the south end of Lake Michigan we see, from left to right, the waves washing sand upon the beach, the wet sand margin, the drift line, the middle beach, the fore dune captured by grass, and in the far background, the trees of later stages. (Photos on this and the next page by Ralph Buchsbaum.)

From the fore dune we see the white caps of the lake (at the far left). The beach lies below the fore dune which is covered sparsely with grass. Behind the grass are the sand cherries and cottonwoods. To the far right are pines and oaks.

Cottonwoods with a few young pines. (South end of Lake Michigan.)

Pines grow in sand after the cottonwoods. (Near Provincetown, Mass.)

Oak growing among pines on Cape Cod near Provincetown, Mass. Just as the cottonwoods prepared the way for pines, so the pines are preparing the soil for the oaks. The oaks will eventually drive out the pines and, in turn, change the soil and provide sufficiently moist conditions so that beeches and maples will drive out the oaks.

Beech and maple forest at the south end of Lake Michigan. This is the *climax* community for this particular region. The big tree at the left is a maple, the one at the right, a beech. The forest is relatively uncluttered.

in the direction of the prevailing winds will sometimes erode away
a large portion of a dune, forming what is known as a "blow-out."
The typical black-oak forest has a number of these blow-outs. The
sand thus exposed will be dried out by the sun and blown by the
wind and may completely cover and finally kill a large part of the
forest. If the sand moves on, it may uncover the standing dead
trees. Such "devastated areas" are quite common. The shifting
sand of the blow-out is ecologically equivalent to the sand of the
middle beach and is slowly brought under control by sand grass in
the manner already described. Often we find small areas in the
black-oak forest where bare sand conditions still prevail and the
animals are those of the cottonwood dune. Surrounding them on all
sides are the dense vegetation and animals of the black-oak forest.
Eventually both become beech and maple forest and the differences
are lost.

We do not have to dig very far into the ground of the beech and
maple forest on sand to find the sandy substratum and to realize
that although we have come many miles from the shore of the present lake, this sand was once a part of the lake. Its present situation is due partly to the inland migration of the dunes and partly to
the constant filling in of the lake by the sand deposited at the south
end. This peculiar situation has left us a series of land communities arranged according to age, with the oldest farthest from the
lake and the youngest on its very shore. In a few hours walk we
have seen the evolution of a forest from dry sand to damp humus.
We could see the same changes by camping for many hundreds of
years on the middle beach of today. At the end of our stay the lake
would have receded and we would find ourselves several miles
from the shore of the lake and surrounded by a beech and maple
forest.

CLIMAX, SUBCLIMAX, DISCLIMAX

On the almost lifeless sandy beach of the dune succession there
was a maximum loss of the energy that reached it from the sun.
Nearly all the solar radiation was promptly changed to heat and
then lost. In the gradual transformation from sandy waste to beech
and maple forest, each stage made increasingly efficient use of the
sun's energy, trapping and converting more of it into living sub-
stance, into increased fertility and stability of the soil, and into a
maximum use and conservation of available water. In other words,
the maturing of the community tends to reduce the stress to a mini-

A moving dune slowly smothers a beech and maple forest. The unstable sand is itself being invaded by grass which slows down the creeping mound of sand. Cape Cod near Provincetown, Mass.

A blowout, leaving behind the remains of a previous forest; some of the trunks are still standing. This bare area will now begin the dune succession all over again--part of this is already started. Notice the partly captured mounds to the left covered with cottonwoods. (South end of Lake Michigan.)
(Photo by R. B.)

mum.

The kind of vegetation that makes the most efficient use of all
the resources of a community, that reduces the stress to the point
where the community is indefinitely self-maintaining, is called the
climax vegetation. The total community it supports is the climax
community. Since a true climax is under the control of climate
(and not some special interference by man or by something like
recurring fire) we also call it a climatic climax. The same kind
of community, the oak-hickory forest, may be only a seral stage
leading towards the beech-maple forest, as we saw in the dune
succession. Or, east of the Alleghenies, in a somewhat drier climate that cannot support the more water-demanding beeches and
maples, the oak-hickory forest can be the climax.

In southwestern Michigan, however, the same climate will support beech-maple forest in'favorable soils, oak-hickory on coarser
soils which hold less water. In such a case beech-maple is the
climax, oak-hickory a preclimax stage, which may go on for a
very long time. Conversely, where beech-maple forest is the
climatic climax but is infiltrated locally (due to special conditions
that conserve moisture) by many hemlocks, the hemlock area is
called postclimax. Temperate climatic areas grade off on their
borders into other climates, cooler and wetter to the north, warmer and drier to the south, so we expect beech-maple forests to
yield gradually to postclimax hemlocks and white pines and bass-
woods along northern borders and to preclimax oaks and hickories
along drier margins.

When an area seems to have been arrested, for a very long time,
at the stage immediately preceeding the climax stage for the particular climate, we call it a subclimax. This may result from extremely slow development to climax as when pine trees in north-
temperate climates yield only very slowly to hardwoods. Or, it
may be due, as in our eastern coastal states to man's interference
and to recurring fires that favor the subclimax pines over the
hardwood forests that should be the climatic climax. Subclimax
may also result because rabbits or men destroy the tree seedlings
that would convert grassland to forest.

Many of our Appalachian oak forests were once oak-chestnut forests, but the chestnuts succumbed to a blight. When the climax is
changed because of the destruction of a species or its displacement

by other species, as a result of some disturbance, the modified climax is called a <u>disclimax</u>. On our great western plains the short grasses that were once thought to be climax for the region are now considered a disclimax resulting from overgrazing and drought. Wherever man-made conditions are eliminated by some sort of special protection, the original taller species of grasses return.

To tell whether the vegetation in any extensive area is preclimax, postclimax, climax, subclimax, or disclimax, is not always an easy matter. It requires knowledge of what the original vegetation of the region was before the arrival of men and their axes or their cattle. It also requires considerable knowledge of the plants of the whole climatic region to determine which dominants form the moistest community that can maintain itself indefinitely in the regional climate. But how do we know the regional climate? We usually define it by the kinds of plants that grow there. If the original trees have all been cut, the original grasses mostly displaced by ones that stand grazing better, or the soil recently covered by lava that sets back plant growth in almost any climate, the climax may not be easy to determine.

To tell whether the oak-hickory woods near your home is climax or merely a seral stage leading towards a beech-maple forest you would have to know whether there are stands of beech-maple reasonably close. If so, you could guess that the wetter beech-maple is the climax for your region. If not, you would have to know whether beech-maple or oak-hickory forest was the original vegetation in your area.

Throughout the story of succession given here we have emphasized the role of the plants and animals in changing the community. Physiographic factors, like erosion, have been omitted for the sake of simplicity. Yet all communities are affected by the work of wind, running water, and other physical forces. In certain habitats the effect of physical forces may be so great that plants and animals can do little to change the environment even over considerable periods of time. Such situations are the bare rocky areas of the arctic or the bare lava surfaces at high altitude in the west. In a rapidly running stream the erosion of the stream bed is more important than are the life-activities of the plants and animals in changing the community. The plants and animals merely follow into appropriate habitats as they form under the action of physiographic forces.

Finally, we see that it is the exception rather than the rule for any community to remain always the same. Even though climate does not change, there will be local upsets: earthquakes, volcanic eruptions, floods, forest fires, droughts, landslides, and the unknowing or purposeful meddling of man. It makes little difference whether we begin with a pond or the dry sand dune; both may eventually become a beech-maple forest. In other words, the various forces acting all <u>tend</u> toward some stable equilibrium which is the fullest expression of plant and animal life possible for any particular region. The extremely wet places keep getting drier and the extremely dry places keep getting wetter, and when they have converged to an intermediate type (climax) a catastrophe of some sort will eventually intervene to start things all over again.

SUCCESSION IN HUMAN COMMUNITIES

The ecologist likes to speculate about succession in human communities, though this is really the province of the human ecologist who is generally a sociologist. The old European cities have streets and buildings which have not changed significantly for hundreds of years. They are in a relatively stable or climax stage. Occasional wars muss things up, create bare areas; the cities are rebuilt and the climax is quickly reached again. Our much younger, constantly expanding cities are still in various seral stages. Once-prosperous neighborhoods "run down" and are occupied successively by poorer and poorer people. Accompanying the degeneration of the physical environment, the buildings, and the changes in the human population, are the cockroaches, mice, bedbugs, lice, and rats which come to occupy the community. When conditions get very bad, the land becomes more undesirable as a place to live, and factories replace homes. The people move to other neighborhoods. Eventually some sort of climax stage can be expected—what it will be depends not only upon physical factors and biotic factors, but also upon the complex interrelations of the elements of human society.

A highway facilitates transportation; ecologically, it is a permanent *bare area* and must be managed as such. Over 20 million acres are occupied by roads in the U.S. (Photos on this page by University of Pittsburgh Photo Service.)

Succession in great cities like Pittsburgh involves the changing of neighborhoods from residence of one grade to another, to commercial or industrial use, and, perhaps, back to residential purposes. Power tools erode the earth creating bare areas on which to build anew.

Left, old, substandard housing giving way to industry and a new highway.

Below, left, sub-standard housing in contrast with a new housing development.

Below, stages in succession as the "Point" undergoes renovation.

Chapter 7

THE DISTRIBUTION OF PLANTS AND ANIMALS

THE POLAR BEAR is not found in both polar regions, and should be called the arctic white bear. It was named long before anyone had seen the Antarctic Continent and reported that no bears or other land mammals were to be found there.

Polar bears do survive the heat of temperate summers when we keep them in zoos and feed them plenty of fish. So it is not merely our unpleasant climate which ordinarily keeps polar bears from coming down to join us. In the temperate zone we also lack other arctic comforts of life, such as year-around drift ice from which polar bears do much of their feeding on seals. They also eat young walruses, fish, the carrion of stranded whales, the eggs of birds, and some seaweed. In summer they go inland for some distance and at such times show the omnivorous habits of all bears by eating berries, grasses, and small mammals. Most of the year the polar bear roams on the ice-pack, and has been seen swimming vigorously many miles from land. If we mark on a map the southernmost limits of the distribution of polar bears it follows closely the southernmost boundary of arctic drift ice. Both coincide with the 0° C. isothermal line which connects all points on a map sharing an average yearly temperature of 0°C. (32°F.).

The cold interior of Canada or Greenland furnish low temperature and substrate, but lack the marine food of the polar bear. Cold waters and abundant marine food extend south of the drift ice, but the polar bear must have a combination of all three things: cold water, marine food, and drift ice. If any one is missing, so is the bear.

All this seems to explain the distribution of the polar bear until we ask why it is absent from antarctic shores where there are all of the appropriate factors. The easy answer is that it has never succeeded in getting there from its site of origin in the northern hemisphere. The Antarctic Continent is so far from any

other great land mass that only truly aquatic mammals like seals
and whales have been able to come near. Some species of whales
are found in both polar seas and apparently cross equatorial waters
to get there. The polar bear may be ecologically suited to live on
antarctic shores, but unless he gets a lift from man he is not likely ever to reach there under present conditions.

That world conditions have changed markedly in the past, and
are in fact doing so today, we know from a great many lines of
evidence. The Chicago region, for example, was referred to in
the description of pond succession as being a transition area between forest and prairie. Yet is is underlain by rocks full of coral
and other invertebrate fossils that tell of the time when the region
was covered by a tropical sea. The coal mines of Illinois (and of
Pennsylvania) are a heritage from the sub-tropical forests of another period of the past. It was not long ago, geologically speaking, that the Chicago region was a treeless tundra like that of
present-day arctic areas.

Geologists furnish us evidence of the sinking and emergence of
coastal areas and of land bridges as sea levels rise and fall, of
the upthrusting of great mountain ranges and their subsequent
erosion, of the earth-scouring action of massive ice sheets. We
use these physical evidences to explain how plants and animals
have changed to meet a changing environment. We also bolster the
geological argument with evidences of plant and animal change as
revealed by an extensive fossil record. Plants and animals are
very sensitive indicators of climatic change, and their fossils are
often much more useful for filling in exact details of climatic
change than any kind of physical evidence that can be gathered.

The "growth rings" of woody cells laid down in the trunks of
trees are larger in rainy seasons of the year than in dry seasons.
In the American Southwest, where the rainfall is all concentrated
into a short rainy season and the rest of the year has little or no
rainfall the "annual rings" are more clearly marked than in temperate regions that have more distributed rainfall. Scientists have
been able to construct year-by-year records of the rainfall pattern
of the southwestern region during many thousands of years.

In many parts of the world botanists are making analyses of the
stratified layers of microscopic plant pollen grains that have ac-
cumulated for thousands of years in certain places. Peat bogs and

peat lake bottoms are usually studied, because conditions in such habitat favor preservation of the hard-walled pollen grains and because peat and lake deposits can be sampled by bringing up long cores in which the layers are relatively undisturbed. Microscopic examination of these cores permits identification of the species of plants and estimation of their relative abundance. Pollen analysis has thus been used as a technique for studying long-term changes in vegetation, complete with tree and other plant species, that have accompanied long-time climatic changes. The "pollen profiles" of the past bear out the patterns of community succession that ecologists have deduced from studies of modern vegetation.

Comparable with the studies of layer-upon-layer of deposited pollens, are the recent studies of stratified marine deposits. Long cores of sediments are brought up from the ocean floor by special coring devices. When analyzed in an especially sensitive mass spectrometer, devised by Urey, the carbonates in the various layers of microscopic foraminiferan shells show varying proportions of heavy oxygen, O^{18}, to light oxygen, O^{16}. This ratio has been shown to vary with the temperature of the sea in which the carbonate in the shells was laid down. Examination of the various layers of a long core can tell much about changing temperatures in the oceans of the past.

Without all the geological and biological evidences of past change in topography and in climate we would indeed be hard put to explain many of the peculiarities of plant and animal distribution we find today. For example, a whole series of arctic plants and small animals occur near the top of Mt. Washington in New Hampshire. Their nearest relatives are to be found in Labrador and Greenland. Plant and animal species do not show this kind of discontinuous distribution without reason. The explanation comes from geological evidences of tremendous glacial scouring of large areas of the northern half of the country, plus other evidences of a glacial past, such as soils overlain by glacial debris, old glacial moraines, etc. Among the biological remnants of glacial ages are the bones of the musk ox in West Virginia, Kentucky, Arkansas, and Utah. The bones of the caribou have been found as far south as Kentucky. From South Carolina west to Texas we find fossil remains of pines, spruces, and larches of species that are now characteristic of northern Canada and Labrador. Piecing together all the parts of this scientific puzzle, we have arrived at a story, well supported

Piston coring tube being prepared for lowering into the sea by the crew of an oceanographic vessel. (Photo by D. M. Owen, Woods Hole Oceanographic Inst.)

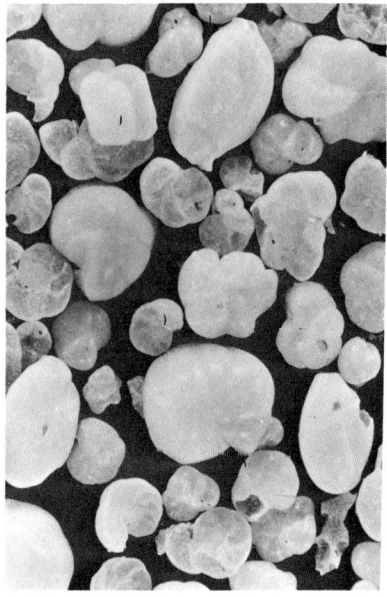

Core from the bottom of the ocean is studied in the laboratory. (Photo by Woods Hole Oceanographic Inst.)

Shells of foraminifera are among the organisms studied from cores. These are about the size of sand grains. (RB)

by so many kinds of evidence, of the repeated southward advance and then retreat of great ice sheets. As the ice moved southward, it pushed a cold climate into cool-temperate and into warm-temperate regions. The animals of the north moved with the southward advancing climate, retreated with the receding cold. When the last ice sheet had melted northward, musk oxen and caribou either died and left behind their bones, or they migrated with the receding climate. Many small forms, however, which could live in restricted areas, retreated gradually up the slopes of high mountains, finding at higher and higher successive levels, as the climate gradually warmed, the tundra conditions to which they were adapted. Such are the "glacial relicts" on Mt. Washington.

Certain reptiles found in the "prairie peninsula" and in "prairie islands" in forested country as far east as Ohio are related most closely to reptiles found only in the grasslands to the west. Their discontinuous distribution fits in with other evidences of an eastward extension of the grassland, in the past, into what is now forested region.

Another kind of discontinuous distribution is exemplified by the peripatus, the small caterpillar-like animal that resembles both annelids and arthropods and so is thought to have developed from

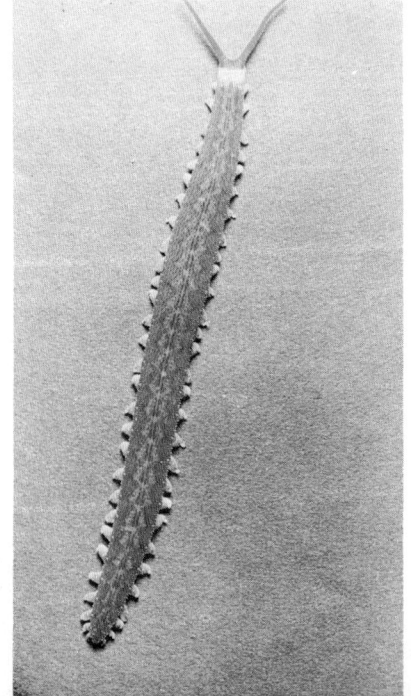

Peripatus lives in moist tropical forests under logs or stones and comes out to feed mostly at night. Largely carnivorous, it preys upon small insects, worms, and other animals. This one, *Macroperipatus geayi*, was photographed in Panama; about 5 inches long; brick-red in color. (R.B.)

the same ancient stock that gave rise to the two phyla. The peripatus occurs in widely separated parts of the world: Australia and
New Zealand, the East Indies, and Southeastern Asia, South Africa
and tropical America including the West Indies. Yet in all these
places it is found only locally and under very moist and restricted
conditions. No animal of such restricted habits could have spread
so widely unless it was at one time a much more successful and
widespread form. It has probably become extinct over much of its
geographic range and at present is gradually disappearing.

The listing and mapping of the distribution of modern plants
and animals (biogeography) and of the plants and animals of the
past (paleobiogeography) are two branches of ecology which are
grown so large as to be separate sciences. They can only be al-
luded to, not really discussed, in this chapter. They contribute
importantly to our understanding of ecology and ecologists contribute to these sciences by their careful analyses, both observa-
tional and experimental, of the physical and biotic factors that
help to explain how animals are able to disperse as they do, and
what are the barriers that keep them from reaching many habitats
to which they seem ecologically suited.

THE DISPERSION OF PLANTS AND ANIMALS

The present distribution of any plant or animal is a result of its
capacity to multiply and then to spread, within the limitations of
time and of physical and biotic conditions, to all of the regions in
the world to which it is ecologically suited. This refers not just to
capacity to spread at the present time but also at any time in the
past during which the group existed. The Asiatic camel, for example, if introduced by man into our American southwestern deserts,
does very well. Then why does it not naturally live there? The
answer is that the Asiatic camel cannot cross either the water
barrier of Bering Strait or the climatic barriers of northeastern
Asia, Alaska, and northern Canada to reach the American desert.
Fossil evidence, however, tells another story. It shows that the
camel family originated on the dry plains of western North America, lived there until relatively recent times, and then became extinct in the site of its origin. This helps to explain how the closest
relatives of the Asiatic camel, the other camel-like mammals of
South America, reached that southern continent. The original
camel ancestral stock spread both south over the Panamanian land
bridge and north over the Bering Sea land-bridge, at a time when

such a bridge existed. After the plains camels died out in North
America, the mountain and plains-adapted camels of South America continued to thrive, as did the camels of Asia, which by then
had become desert-adapted. So the present geographic range of a
group is only a part of a larger story. It may or may not tell where
the group originated; it may or may not indicate the particular
habitat to which the group was adapted in the past. The barriers
that limit present distribution are not necessarily those that oper-
ated in the past.

With all this in mind, we may briefly examine some of the
means by which organisms are dispersed.

Plants spread, with rare exceptions, by passive dispersal of
their spores, seeds, or fruits. Spores are light and easily carried
by winds or by moving water, or in the mud that clings to the feet
of birds. Seeds may have special expansions, like the "wings" of
maple seeds, that make them easily carried by the wind. Or seeds
may be covered with curved hooks, like those of the cocklebur,
which catch on the wooly fur of mammals or the clothing of man.
Seeds and fruits are eaten, especially by birds and mammals,
and then deposited in excreta beyond the area from which they
were taken.

Animals also may be spread passively, either as adults or as
eggs or larvas. The spores of protozoans, and the cysts of small
multicellular animals, can be blown by wind, moved by water, or
carried in mud on the feet of birds or of mammals. Small animals
are transported on rafts of floating wood or of sod that have broken
loose from a shore. The polar bear travels on floating ice for
hundreds of miles. Ordinarily, however, animals disperse by their
own active efforts in creeping, running, hopping, flying, or swimming.

The rate of spread varies greatly. The natural spread of vegetation over Sweden (up to lat. 61°) as the last great ice sheet retreated was almost as fast as the rising temperature allowed—400
miles in 3,000 years. This is faster than one-tenth of a mile a
year, and very rapid indeed for plants that are rooted where they
grow and must spread passively.

The maximum rate of spread for trees of the walnut family,
which depend mostly on squirrels to carry their nuts, has been

estimated at about a mile in 1,000 years. Actually, conditions for
spread are not always favorable, and a forest of walnuts and hickories that could have spread from the site of its earliest fossils
in Alaska to that of its earliest fossils in Oregon, a distance of
1,800 miles, in two million years, may have taken 10 million years,
according to a leading botanist. An ecologist might be tempted to
put the figure much lower on the ground that spread does not always
occur step by step; sometimes it takes place by lucky jumps. The
tree-dwelling snail, Liguus, is found in seemingly endless variety
in Cuba, Haiti, and southern Florida. On the island of Cuba certain varieties inhabit specific, very limited areas, and these same
varieties are found only in specific spots in southern Florida. When
all these "twin-areas," both in Cuba and in Florida, are marked
on a map they are shown to lie directly on the path of historic hur-
ricanes. It is thought that the young snails may have been carried
by the winds while attached firmly to leaves.

Aside from his natural role as a mammalian disperser of spores
and seeds and fruits, man now consciously disperses animals. He
introduced the English sparrow into Nova Scotia and New York City,
and this alien bird, lacking the natural enemies that held it in check
across the Atlantic, covered all of settled North America in about
50 years.

In modern times man disperses many flying insects that enter
airplanes, and many aquatic animals that encrust ships. After the
last world war, it was found that snails were being brought to the
United States in the caked mud under motor vehicles. Every ship
returning from the Pacific Area had to have its cargos of plane
and motor vehicle tires sprayed with insecticides. Half of the tires
aboard had water in them, and in most of these tropical mosquitoes
were breeding.

Ease of dispersion varies with size and with special adaptations.
The same species of microscopic protozoans, whose minute cysts
weigh little, occur on every continent. On the other hand, no land
mammal ever reached the Hawaiian Islands on its own, though
these isolated islands are estimated to be five million years old.
During that long period many other animals and plants did reach
the islands, and in their long stay there evolved into a large number of new species unique to Hawaii:

COLONIZATION OF HAWAII BY PLANTS AND ANIMALS

Data from Simpson after Zimmerman, 1948.

Groups	Estimated number of ancestral species which invaded islands in the past	Native species of Hawaii today
Land Mammals	0	0
Land Birds	15	70
Insects	250	3722
Land Snails	25	1061
Seed Plants	270	1633

From this list it can be seen that winged animals have a tremendous advantage, though small animals like snails that can be carried by wind or can cling to floating vegetation also do well. Spiders are extremely light and often spin threads that are readily caught up by the wind. They are well represented in the Hawaiian Islands, and we may recall that a spider spinning its web was said to be the first living organism noted on Krakatoa nine months after the explosion. On that devastated island the ferns, which have exceptionally light spores, constituted 40% of the vegetation three years after the explosion. Ordinarily, a tropical island would have no more than 10% to 20% of its stable vegetation as ferns. Orchid seeds are almost as light as the spores of ferns, and thirteen years after the eruption, four species of orchids were flowering. This was in spite of the fact that Krakatoa still lacked the rich soil and the trees that orchids usually require. An organism that is readily dispersed has repeated opportunities to colonize a new area, and if it fails to find suitable conditions one time may do so on its next arrival.

BARRIERS TO DISPERSION

Since plants cannot move into the shade when the sun is high in the sky, or hide in a cave during cold winters, the major barrier to plant dispersion is climate, with temperature limitations more important than moisture differences. Next come edaphic (soil) barriers, since plants require certain soil minerals, specific acid or alkaline conditions, and particular soil textures. Biotic barriers are next in importance and these may be anything from the lack of soil fungi required by the roots of certain trees, to the presence of rabbits that keep nibbling tree seedlings and prevent the advance of a forest. Geographic barriers: rivers or oceans for land plants, continents for marine plants, come last of all.

For animals geographic barriers are certainly of first importance, especially for the larger animals like mammals. Climatic factors are next, and biotic ones last. The sea is the great geographic barrier for land animals, continents for marine animals. Since the seas are all connected, and physical conditions do not vary as much as on land, marine animals are in general more widely distributed than are land animals. A few species like the giant sea turtle and the jellyfish Aurelia, occur in all seas. No large land species are so widely distributed. Every continent has its own characteristic fauna. As we saw for Hawaii, islands that are isolated by great stretches of water have unique faunas.

A narrow strait is not a barrier if it is frozen over for a good part of the year. Between Novaya Zemlya and Spitzbergen (north of Norway), reindeer cross 600 miles of sea when the ice permits.

Small rivers are seldom a barrier to land animals since they may be crossed on floating logs or debris or during low water. Large rivers can be a serious barrier to some animals, though usually not to flying forms. Yet even some birds are stopped by a really large river like the Amazon. The ant bird, Phegosis nigromaculata, lives all along the south bank but not on the north bank.

The same bodies of water that are barriers to land animals are highways of dispersion for aquatic animals. But there are many barriers to dispersion even in the sea. To coastal animals deep water is a barrier. To deep-water forms the higher temperatures of the upper layers of water are a barrier. Changes in salt content (salinity) of sea water blocks dispersion of some animals. Lack of a suitable food keeps marine animals from wandering into certain areas. Even mechanical barriers are known in the sea. The Faroe Channel, which lies north of Scotland between the Faroe Islands and the Shetlands, is divided into a "cold" and a "warm" area by a submarine ridge that rises from the ocean floor to within 1800 feet of the surface. On the Atlantic side deep trawling brings up all the typical deep sea forms of the Atlantic. But on the Norwegian side of the ridge deep trawling brings up an entirely different set of species. In the surface waters above the ridge there is a large admixture of typical surface Atlantic forms with the cold water species of the Norwegian Sea. Apparently the ridge (called the Wyville Thomson Ridge for the man who first noted the difference

in the two faunas) effectively separates the deep water animals of
the Atlantic and Norwegian sea basins.

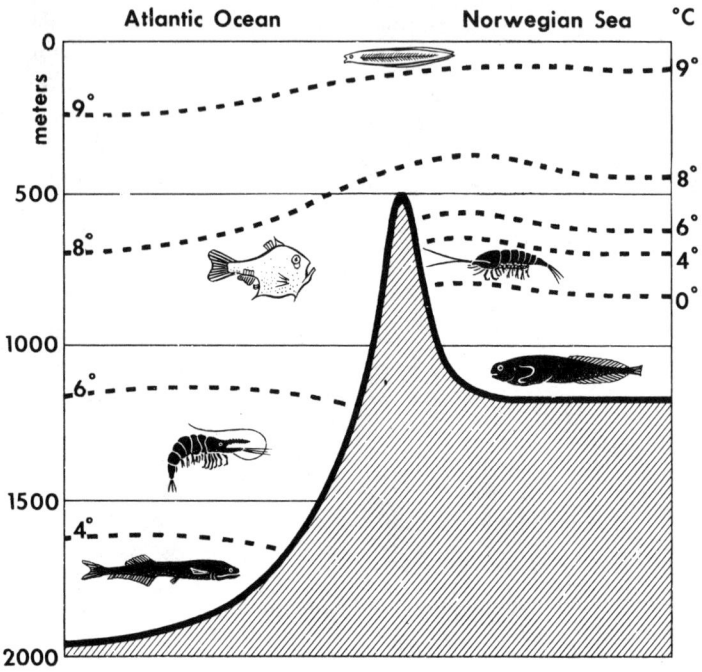

The Wyville Thomson Ridge shuts out the Atlantic deep-water fauna from the Norwegian Sea. Black fishes and red crustaceans typical of cold deep waters occur on
both sides, but are of different species. The Atlantic fish shown in the 6-8°C. (43-
47°F.) layer of water that crosses the top of the ridge, is sometimes found on the
Norwegian side. Animals of the surface waters (like the eel larva shown at the top)
are found on both sides. (Based on Murray and Hjort.)

Marine animals that have floating (planktonic) larvas usually
cannot invade rivers because the larvas keep getting washed downstream and out to sea. As a result, whole groups of marine animals
are missing from fresh waters. This, together with physiological
requirements has kept the echinoderms from fresh water. The
coelenterates are all but unrepresented in fresh water. One of the
few representatives, the fresh water hydra, has eliminated any
free–floating stage.

The same Panamanian land bridge that now makes possible an
exchange of North American and South American land forms, is a
barrier to exchange of marine forms between Atlantic and Pacific

Oceans in the Panama region. Nevertheless, we find a very great number of similar species on the two sides of the isthmus. The spiny lobster, of the southern waters of the United States, for example, is more closely related to a spiny lobster in Pacific waters, than it is to the lobsters of more northerly Atlantic shores. This same kinship between many "paired species" of fishes and invertebrates on the two coasts of Central America suggests that the waters in which they live were once continuous. And there is much geological and fossil evidence to support the theory that there was once a broad connection between the two oceans at the Central American level. The submergence of the land bridge in the Panama region during a very long period also helps to explain why the large North American mammals are more like those of Eurasia than they are like those of South America.

For land animals climatic factors come second only to geographic barriers. Every desert is a barrier to forest-dwellers, every forest a barrier to many prairie animals. A mountain can be both a geographic and a climatic barrier, since an animal that tried to climb one would have to adapt to a whole series of increasingly colder climates as it moved to higher and higher levels. Animals that live on mountains or on high plateaus may find valleys impassable.

Kaibab squirrel. The Grand Canyon of the Colorado now separates two species of squirrels that were once part of the same population of the plateau into which the canyon has been cut. On the north rim lives this white-tailed squirrel; on the south rim lives another species, isolated from its relatives. Many of the other animal populations of the cool coniferous forest of the north rim are separated from those of the south rim not only by the gap of the canyon and the swift river below but by the desert climate at the bottom of the canyon. (Photo, R.B.)

Biotic factors, such as suitable food, keep many specialized animals from moving into or through communities other than their own. A desert is both a climatic and a biotic barrier to a forest-dweller. The dependence of some animals on specific kinds of plant food makes such animals subject to any barrier that affects their plant food.

Some animals are limited in their distribution by particular competitors, others by their predators. The marine mussel, My-tilus, can live in still waters but is not usually found there. It lives mostly on rocky shores which are subject to enough wave action to eliminate its competitors. In some areas it seems to be barred by its enemy, the starfish; in other areas it disappears when the octopus is on the increase.

Accidental dispersion of marine animals by man is illustrated by the barnacles on this piece of driftwood at the edge of Salton Sea. The small inland "sea" was formed about fifty years ago during the construction of an irrigation project when water broke from the canal into the Salton sink in the Imperial valley in southern California. This low land, more than 200 feet below sea level, quickly filled with water from the Colorado River. The breach was soon closed and the water has been evaporating ever since. It is now about as salty as the ocean. In recent years marine aircraft have been shuttling between Salton Sea and the Pacific Ocean. The result has been the introduction of barnacles and a marine annelid worm into Salton Sea. (Photo by R. B.).

Chapter 8

BIOMES OF THE WORLD

IN PLANNING a vacation we all become amateur ecologists and biogeographers. Specific demands as to temperature, duration of sunshine, amount of rainfall, and the presence or absence of particular kinds of plants and animals all have a part in the planning. But rare is the vacationer who really studies geography or ecology texts before setting off. More likely references are travel brochures and the enthusiastic testimony of friends, so it is not surprising that we often go wrong and arrive in Maine the week the black flies are biting most, or in California during the rainiest month of the year. On the whole, though, most of us do fairly well merely by remembering to go south in winter and north in summer!

The latitudinal zonation of the earth's climates results from a temperature gradient based on differences in the amount of solar energy that reaches successive regions from the equator polewards in both directions. The quantity of energy received varies with the angle at which the sun's rays strike the surface of the earth at different latitudes and at different times in the solar cycle. Were the land surface today relatively homogeneous and smooth, the terrestrial climatic zones, and the vegetation they support, would come much closer to being neat latitudinal concentric belts crossing the continents. Geography and ecology would be simpler sciences, weather more predictable, and this chapter much shorter.

There have been long periods in the earth's history when the continents were relatively level, following long periods of erosion. The enlarged seas occupied much of what is now land surface, greatly moderating land climates. Ocean currents circulated freely from the tropics to the arctic seas, so that heat and precipitation were well distributed over the whole globe. During some periods practically all of the land surfaces were either tropical or temperate in climate, with enough moisture to support rich forests over a great part of the continents. Tropical rain forests, now limited to an equatorial belt, extended far beyond their present boundaries. Warm temperate forests pushed northward to the region of the Arctic Sea. Fossils of magnolias, palms and figs in Alaska, and of redwood trees in Alaska and in Spitzbergen, attest to the spread of mild climate at certain times.

During the periods when a land bridge provided solid footing across the Bering Sea from Alaska to northeastern Siberia, and climate permitted, the plants and animals of North America and of Asia were liberally exchanged. At one period of mild climate a moist temperate forest occupied most of Canada and stretched across Eurasia at comparable latitudes. Among the oaks, maples, beeches, chestnuts, and birches—all trees familiar to us in our eastern forests—there were also many members of the redwood family. The redwoods probably originated somewhere near Alaska and from there spread throughout Canadian forests and over the Bering land bridge into Asia and then Europe along with other temperate forest trees that originated in arctic regions during this long warm period in earth history.

At times of active interchange between Asia and North America, South America was an island continent, cut off from North America by a submergence of the Panamanian region. So South America has a fauna which is strikingly different from that of other continents. Interchange with North America did take place when the Panamanian bridge was functioning, as it does today, but on a much smaller scale. Most of the traffic was from north to south, but we did receive from South America the stocks from which are descended our armadillo and our porcupine.

The great similarity between European and North American plants and animals, especially the conspicuous land mammals, makes English and European fairy tales and novels sound familiar. References to the many temperate forest trees, to rabbits, deer, foxes, bears, and wolves, evoke clear images. This would hardly be true of a South American tale that spoke of jaguarundis and coatis, tamanduas and tayras. While it is true that this last group of animals is typical of the South American tropical forest, an ecological setting which we do not have in our temperate climate, the European and North American similarity is not just a matter of similar climate. It is largely a result of interchange of flora and fauna through Asia. Only a small part of the similarity is due to interchange across the Atlantic. For large mammals Atlantic

interchange has probably played no role at all. Like the other
large mammals, man came to North America only by the Asiatic
route until he finally learned to build boats that were seaworthy
enough to cross the Atlantic.

After the last great erosional cycle there came a new cycle of
earth history, a period in which great mountain chains were thrust
up and continental areas enlarged. Inland seas were drained, and
land masses interfered so much with oceanic circulation that less
heat was carried to the Arctic Sea. That great body of increasingly cold water came to support vast ice floes. The mountain ridges
interfered with the formerly even sweep of the wind belts across
the continents, causing excessive moisture to be precipitated on
the windward slopes and resulting in dry "rain shadows" on the
leeward slopes. Since most of the mountain chains tend to occur
around the edges of continents, the temperature-regulating influence of wind masses sweeping in from the oceans was cut off from
the interiors of the continents, producing both hot and cold deserts
and grasslands where forests formerly grew.

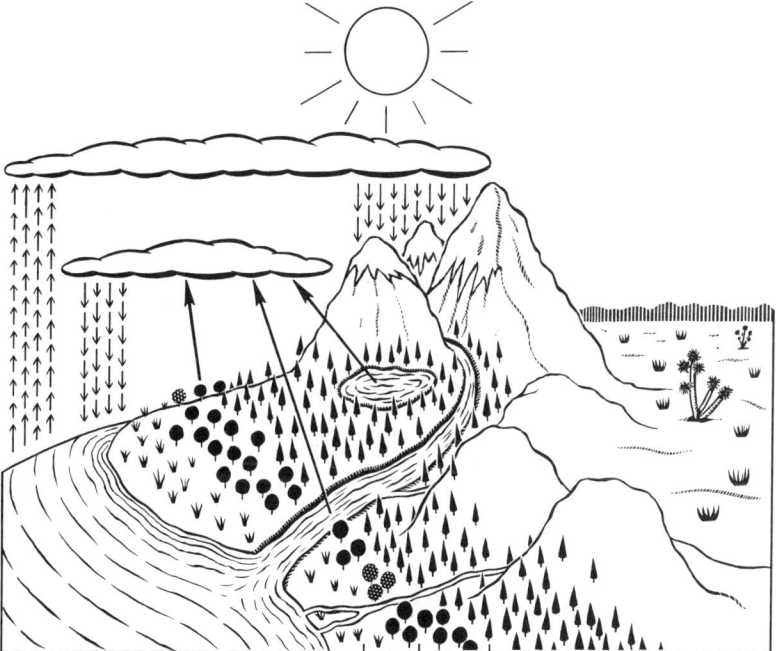

Mountains produce "rain shadows." Diagram of a portion of the West Coast
in California showing the water cycle west of the mountains and the desert
east of the mountains.

At temperate latitudes such continental areas have very hot summers and very cold winters, as those who live in the central part of the United States well know. In the geologically recent past, this cycle produced the many ice ages, and we are now still in the erosional cycle that is following the last great period of mountain building.

When the climates began to cool, the temperate forests of Canada and of Siberia retreated southward, replacing the tropical trees which were also being pushed equatorward by decreasing temperature. The up-thrusting of the Rocky Mountains had created, in the middle of the American Continent, a great arid region where

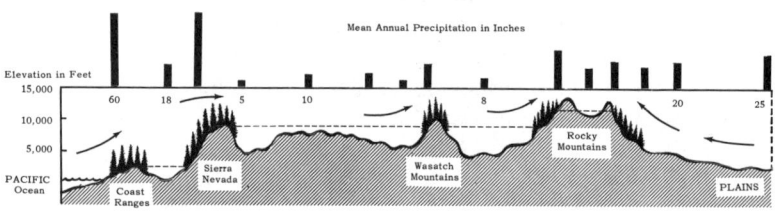

Moist air currents in relation to altitude affects the distribution of forests in the West and in the East as shown in the diagram across the United States along the thirty-ninth parallel of north latitude.

once there had been tropical and sub-tropical seas, swamps, and forests. As the temperate forest of Canada reached the dry mid-continental plains it became divided and moved down on either side of the plains. The eastern portion became the forested eastern third of the United States. The western portion occupied much of the area from the Rocky Mountains to the West Coast. When the Sierra Nevada mountains were again uplifted, and when the Cascade Mountains arose, creating another "rain shadow" on their eastern slopes, another arid intermountain region was added, all but eliminating the western temperate forest except along a narrow coastal strip. Thus the United States has no great latitudinal vegetational zones that reflect the decreasing temperature from north to south. Instead the north-south mountain ranges have broken the country up into an east-west series of vertical vegetational zones based primarily on rainfall differences. The two coasts are forested. The great central portion of the continent, cut off from moisture-laden clouds by mountain ranges and by remoteness from the oceans, is either desert or grassland.

The coastal temperate forests of Oregon and Washington are today mostly evergreen forests, but they do have enough oaks, maples, and birches to remind us of the long-past connection with the eastern deciduous forests. The redwoods of past times have all disappeared from the eastern part of the continent, where they ranged even to New Jersey. Now the "coastal redwoods" (Sequoia sempervirens) are confined to coastal California and Oregon and the related redwood "big trees" (Sequoiadendron giganteum) to the unglaciated parts of the western slope of the Sierra Nevada range in California.

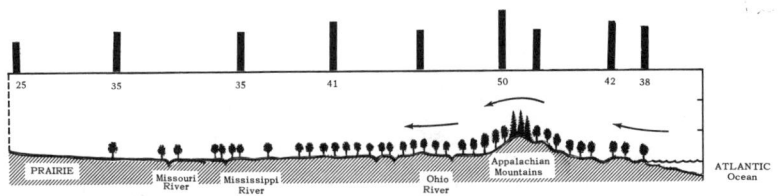

(From U. S. Department of Agriculture Yearbook, *Climate and Man*, 1941.)

These modern redwoods are not the ones which ranged so widely in times past. The ancestral genus, Metasequoia, which we call "the dawn redwood" is the one which has left abundant fossil remains in Oregon and in Alaska and elsewhere. Today it survives only in certain valleys in Central China, where it grows among oaks, beeches, birches, and walnuts, and other remnants of the once-great arctic temperate forests. These forest relicts in China look almost exactly as did the forests of Oregon some tens of millions of years ago. Judging from the leaves of the Chinese trees, and the leaf fossils of Oregon, there has been no great change in these trees during millions of years.

All of this helps to explain why the great ecological areas of the world, in our cycle of world history, are not everywhere arranged in simple latitudinal belts. We begin to understand why ecologically equivalent forests or grasslands in South America and in North America have quite different kinds of animals, whereas comparable

habitats in North America and in Europe support a more similar
fauna. Geographic position, absolute distance, and the past history of climates, and of intercontinental connections, in relation
to the time of origin of various organisms, have restricted or
promoted plant and animal similarities in similar habitats. African giraffes moved about in Eurasia but never got as far as North
American grasslands, much less South American ones. Large
grazing mammals of Asiatic temperate grasslands, such as bison
and sheep, did reach North American but not South American
grasslands. Lacking large grazing herds, the South American
grasslands develped big rodents not found elsewhere.

Petrified forest is proof of vastly changing climate from one which supported the
great *Metasequoia* trees to one in which only small semiarid plants can live.

Nevertheless, and especially where geographic barriers are
not so extreme as between South America and Africa, plants and
animals that live in similar habitats tend to resemble each other
more than they do organisms of dissimilar habitats. The oaks,
tree squirrels, and owls of a Pennsylvania oak forest are more
like the plants and animals of an oak forest in England than they

are like the grasses and meadowlarks and ground squirrels of the
nearby Illinois prairie. So it is quite possible to generalize about
forests all over the world, or about grasslands wherever they may
occur. The plants and animals may or may not be taxonomically
related, but the plants will have similar adjustments to climate
and will look superficially similar; the animals will have many
external resemblances in form and in coloring, in habits of feeding and breeding. They will occupy most of the same niches because similar opportunities are open to them.

LAND BIOMES

   If we map the world in terms of great ecological regions we
see that these have little relationship to the political world map,
but that they do correspond closely with a map of world climates.

   The reason for this is easy to see. The climax plant communities of the world are climatically controlled, and on the land it is
the plants that provide the ecological setting for animal life by furnishing the base of their food supply and shelter from the extremes
of climate. The major world-wide ecological units or biomes that
we usually recognize—the tundra, the coniferous forest, the temperate forest, the grassland, the desert, the tropical forest—all
are named for a general type of climax vegetation which has main-
tained itself in any region for a very long time.

   The temperate forest biome of eastern North America includes
a number of climax deciduous forests, such as beech-maple, oak-
hickory, or oak-chestnut, but all are climax temperate forests.
This biome further includes all seral stages leading to these various forest climaxes, be they dry sand dunes or lakes. Also in-
cluded, of course, are all the animals of the constituent communities of the biome.

   The temperate forest biome provides a certain excess of precipitation over evaporation that enables it to support trees. Where
this ratio falls below a certain point the forest grades into a grassland biome. With increased aridity, grassland grades very gradually into a desert biome. At its northern temperature limits, the
temperate forest grades into the coniferous forest biome.

   The biomes of the world, being climatically controlled, are
arranged in a general way in concentric belts that follow the lati-

tudinal temperature gradient. But many other factors enter into producing the patchwork of biomes we see when we look at a map of the world's vegetative regions.

Only at the top of the globe are there circumpolar biomes. Surrounding the permanent snow and ice of the north polar cap is a ring of tundra biome, narrow over Europe, wider over the top of Eurasia and North America, where at its eastern edge it dips down along the coast of Labrador into Newfoundland. Below the tundra and stretching in a wide belt, 400 to 800 miles wide, across Canada and across Eurasia, is the coniferous forest biome. Like the tundra, the coniferous forest is essentially a single biotic community, interrupted only by Bering Strait and by the north Atlantic Ocean and having very similar plants and animals throughout.

Southward from the coniferous forest biome, all the latitudinal temperature zones are to a lesser or greater extent broken up into several biomes by wind currents, ocean currents, mountain barriers, soil differences, etc.

Along its southern margins the Canadian coniferous forest adjoins every major biome that can be supported by temperate climate. Crossing the United States from east to west, there are vertical vegetational zones of deciduous forest occupying approximately the eastern third of the country, a narrower zone of humid grassland or prairie, a wide zone of dry grassland which we call our western plains, desert plains with sparse grass and shrubs, and a narrow coastal strip of temperate rain forest. Where mountain ranges come down into any of these biomes they carry, on their higher levels, extensions of the coniferous forest biome from the north.

Similar east-west rainfall gradients can be seen across other continents at temperate latitudes—in Asia by going west from Korea, and in Africa and Australia by going westward from the southeast coasts.

In Europe the deciduous temperate forest pushes so far northward that southern Scandinavia has deciduous forest at latitudes at which Labrador, on the east coast of Canada, has tundra. The southern tip of England supports palms through the winter months that bring deep ice and snow to the coast of Labrador. These distortions of the latitudinal temperature gradient are caused by the

cold "Labrador Current" that flows down the east coast of Labrador and by the "Gulf Stream," a warm mass of ocean water that
flows from off our southeast coast up into cold arctic waters. By
shifting England and much of Europe from the coniferous forest
biome to the temperate forest biome, where less human energy
goes into shoveling snow and chopping firewood, the Gulf Stream
has played a major role in shaping Western Civilization. Other
warm and cold ocean currents that determine the boundaries of
whole biomes have also directed the course of human cultural
evolution.

In the southern hemisphere the major continents do not reach
nearly so far into the higher latitudes as in the north. The Antarctic Continent has sparse plant and animal life but nothing that can
be compared with the northern tundra. The coniferous forest biome
is missing altogether, since it would have to occur where there is
only open ocean. Temperate land areas are very restricted and
consist of temperate grassland and forest. The forest is not deciduous, but because of mild climate and high rainfall, as along
the southwest coast of Chile, is a broad-leaved evergreen forest.

Between the temperate zone and the equatorial tropical belt
there is in each hemisphere a zone of subtropical temperatures
and steady winds throughout much of the year. Evaporation is high,
and most of the rainfall comes not in the summer, as in temperate forests and grasslands, but in the cooler or winter season. In
North America this hot dry belt produces the pleasant California
climate near the seacoast, but inland it results in the deserts of
our southwest. In Europe this same zone produces the "Mediterranean Climate" of Spain, southern France and Italy, with hot dry
summers and cooler rainy winters. "Mediterranean vegetation" is
typically evergreen but with small, leathery leaves that withstand
summer drought. Its ecological equivalents are the evergreen oaks
of Texas and the chaparral of California.

Closer to the equator, these two belts produce the great deserts
of the world. In the northern hemisphere the deserts stretch from
the Sahara in North Africa through Arabia and Iran to the great
Mongolian Desert. In the southern hemisphere are the Atacami
and Patagonian deserts of South America, the Kalahari of southern
Africa, and the Great Australian Desert.

The winds that help to create the two desert belts blow steadily

towards the equator, in the convection set up by the constantly
rising warm air of the equatorial belt. The moist, hot tropical
air is cooled as it rises to high altitudes, and the precipitated
moisture comes down most of the year as the heavy rainfall characteristic of the equatorial tropics. The year-around warmth and
rain produce tropical rain forests in Central America, in equatorial Africa and South America, in the East Indies, in India, and in
Malaysia.

Wherever mountains and other physical factors reduce rainfall
in the tropics, forest gives way to tropical grassland or savanna,
a grassy parkland with scattered or clustered trees. The more
humid savannas have many trees, but as rainfall decreases they
grade off into more open grassland and finally into desert.

The world biomes may be
summarized in a diagram which
gives also a simplified picture
of world climates. Temperature
increases from top to bottom;
moisture increases from left to
right. (Modified from: Climate
and Man, U. S. Dept. Agr. , 1941)

Dry Cold				Wet Cold
Perpetual snow and ice				
Tundra				
Coniferous forest				
Desert (arid)	Dry grassland (semiarid)	Prairie grassland (subhumid)	Forests (humid)	Rain forests (wet)

0	16	32	64	128

Dry Hot   Inches rainfall.   Wet Hot

MOUNTAIN ZONATION

A mountain may introduce an island of tundra and coniferous
forest into any biome, even desert or tropical forest. A single
high mountain peak near the equator could support a whole series
of vertical communities ranging from perpetual snow and ice at
the top through tundra and coniferous forest down through deciduous forest to desert or rain forest at its base.

Mountain zonation is a result of change in climate at successive altitudes. At the treeline near the top of Mt. Washburn in Yellowstone Park, Wyoming, dwarfed and windswept conifers give way to treeless area like the tundra which is found at low altitudes only in the Arctic. (Photo by R. B.)

Vegetative zones on San Francisco Peaks, near the south rim of the Grand Canyon in Arizona, from treeless tundra at the top, close to 13,000 feet, to shadscale desert (plant dominant, *Atriplex*) at the base at 5,000 feet, as can be seen in the diagram *below*. Note that zones rise higher on south-facing slopes than on north-facing slopes. (Photo by R.B. Diagram based on Merriam.)

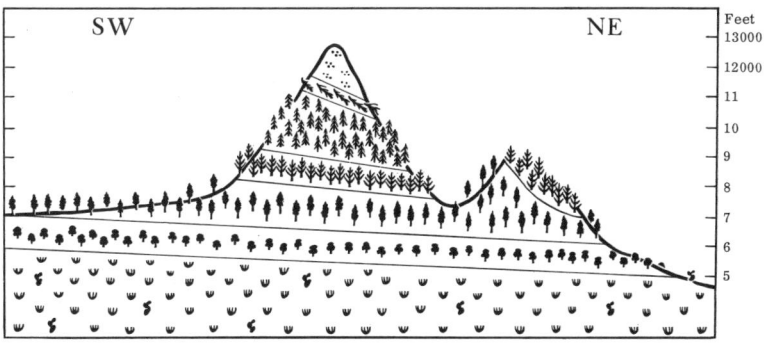

Tundra biome forms ring of varying width around tops of continents. Southern limit is the tree line; northern limit is the Polar Sea. Islands in the Polar Sea, like Baffin Island shown *above* in winter, are ringed with tundra. It occurs wherever melting in summer is sufficient to support mosses and lichens. (Air Force Photo)

Musk ox, shown *below* on tundra in summer, successfully copes with arctic wolves and the cold of winter on the open tundra. Its most formidable enemy is man, who has eliminated the musk ox from parts of its formerly circumpolar range. When winter comes members of the herd huddle close together and feed on lichens under the snow. (Photo by E. F. Keller.)

Forest-tundra ecotone, transition zone between tundra biome and coniferous forest biome, shown *above* **in winter** and *below* **in summer,** about 80 miles south of the tree line on the east shore of Hudson Bay. The ecotone serves as the winter refuge for many tundra animals. As on the tundra, the ground is covered with *Cladonia* lichens, on which tundra caribou (and reindeer in Eurasia) feed. (Photo by John Marr)

Spruces are widely spaced here, but the trees are tall and not deformed as they might be if they were near the edge of their climatic limits. The northern edge of the tree line in this region is not set by temperature but by lack of sufficient soil cover on the bare rock surface left by glaciation. The forest is advancing northward in tundra as fast as soil accumulation permits. (Photo by John Marr)

Antarctic continent is not a distinct biome but an extension of the antarctic marine biome. Man is the only land mammal that can reach antarctic shores. The ship shown *above* is the U. S. *Bear* in the Bay of Whales. (Photos on this page and the next, from Nat. Archives.)

United States Weather Bureau Station obtains records of climate that is less cold than the arctic winter but also less warm than the arctic summer. Constant windiness and cloudiness do not favor plants and animals. There are few seed plants, poorly developed lichens and mosses, few insects, and even these depend partly on organic matter deposited on shore by penguins and other animals that feed entirely from the sea.

Seals come ashore briefly to rest and sun themselves and stay ashore longer during the breeding peri od, but depend upon the sea for food.

Penguins spend much of their time on shore, but like the seals, feed entirely from the sea. The Emperor penguins, shown here, have a special brood pocket between the feet for keeping an egg warm, and are better adapted to withstand cold; they do not have to migrate northward at the start of winter as other penguins do.

Coniferous forest biome, or *taiga* as it is called in Siberia, is characterized by the spruce, and also by firs, trees that can withstand the physiological drought of long winters. The typical large mammal is the **moose,** shown *above,* browsing on conifers in the Canadian forest. The moose browses on the leaves and twigs of trees and in winter on the bark. It also feeds on the herbaceous ground cover and on the aquatic plants of ponds and lakes. It is so closely associated with coniferous forest that ecologists have called this the *spruce-moose* biome. A whole group of birds, the crossbills, whose beaks are adapted to extract seeds from pine cones, are confined to this biome. (Photo by Mildred Buchsbaum.)

Black bear is not confined to coniferous forest but formerly occurred in wooded areas over most of the United States. It is commonly seen by the roadside in the coniferous forest of Yellowstone Park *below,* where coniferous forest biome dips down along mountain ranges into the United States at a latitude that supports grassland to the east. (Photo by R. B.)

![Temperate forest biome photograph]

Temperate forest biome, which occupies the eastern third of the United States, is illustrated *above* by the moist **beech-maple** climax forest of southern Michigan. It includes many kinds of deciduous forests, but all are alike in shedding their broad, thin leaves during the cold winter season. (Photo by R. B.)

Wood frog, *Rana sylvatica,* is a typical "indicator" of moist conditions on the floor of the beech-maple forest. (Photo made near Linesville, Pa. by R.B.)

White-tailed deer (Virginia deer) are typical large mammals of our eastern deciduous forests; large herds live in the oak-hickory forests of Pennsylvania. In summer they browse in open woodland at margins of forest; in winter they shelter deeper in forest, browse on bark and tree twigs. (Photos on this page and the next, courtesy Pennsylvania Game Commission.)

Wild turkey, once widespread as far north as southern Canada, now is restricted to certain states, one of them Pennsylvania. It feeds largely on seeds, nuts, grain, and insects.

Raccoon, *above left,* is a typical animal of well-watered forests, but not of dry ones. **Muskrat,** *upper right,* lives in ponds in large, dome-shaped homes made of reeds. **Opossum,** *lower left,* is the only marsupial in North America. It thrives in a world of placental mammals by staying in trees and occupying a nocturnal niche. Like many arboreal animals it has a prehensile tail. **Cottontail rabbit,** *lower right,* feeds on herbs of forest floor, on tree bark, and on other plants outside the forest in summer; keeps close to forest in winter.

Pymatuning Laboratory for Field Biology of the University of Pittsburgh is loca-
ted on Pymatuning Lake near Linesville, Pa. It offers facilities for research and
also a program of summer classes in the biology of the temperate forest biome,
which there includes nearby lakes, rivers, swamps, meadows, and forests.
(Photos on this page and top of the next page, courtesy Dr. A.C. Tryon,Jr.)

Study of lake biology centers about analysis of the microscopic plankton, which
occurs in Pymatuning Lake in amounts greater than in any other lake. Dr. Rich-
ard Hartman, *below left,* is shown concentrating plankton by the use of a centri-
fuge. **Sampling lake water in winter,** *below right,* through hole in the ice will
yield data useful in studying year-around lake ecology.

Seining for fish in Conneaut Creek is one of the activities of the class in Ichthyology. Rivers and lakes are early seral stages in the successional development of a land community and so are not treated as separate biomes. A small pond in a forest may be only a few hundred years from becoming an indistinguishable part of the forest. A large lake is an extreme example of a seral stage that prevents the climax from being attained for a very long time because of "special soil conditions." Nevertheless, in the very long run all bodies of fresh-water will eventually become completely incorporated into the land biome in which they occur.

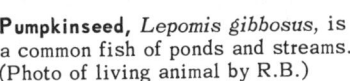

Pumpkinseed, *Lepomis gibbosus,* is a common fish of ponds and streams. (Photo of living animal by R.B.)

Toad on forest floor is an animal that feeds on insects in forest but must return to pond to breed and lay eggs.

Grassland is a region in which perennial grasses are the dominant plants. The eastern portion of the American grassland was originally a mixture of tall grasses, some of them bunch-grasses that grew six feet tall. This *tall grass* or *true prairie* had relatively high moisture requirements for grassland. It persists in protected spots, as in the area in northern Minnesota pictured *above*, but most of the true prairie is now converted to corn and wheat crops. (Photos by R.B.)

Short grass plains extend west to the Rockies. Thought to be a disclimax due to overgrazing and to the increasing aridity that follows, they revert, when protected, to a mixture of short and tall grasses like those of the transition zone with tall grass prairies on their eastern border which runs down through the middle of the Dakotas. *Above*, short grass cattle range in North Dakota.

American bison, once the characteristic large mammal of the American grassland, has been replaced by domestic herds of cattle and sheep. Bison once lived in large herds, as many as 2 million in a herd. Now only a few small herds remain in protected spots. The ones grazing here are in southern Utah in badly overgrazed grassland slowly turning to semiarid desert.

Rodents in large numbers are common to every grassland, whether it be called the American prairie, the Russian steppe, or the Argentinian pampas. In the photo *above*, taken in the Wyoming grassland, two "prairie dogs,"*(Cynomys)* are sitting up on the mounds made next to their burrows, watching out for predators. This habit is seen also in Asian and South American rodents, for in a biome in which visibility is very good and gregariousness characteristic of many groups, the watchfulness of any members serves to alert the whole group. Prairie dogs, mice, ground squirrels, pocket gophers, and rabbits provide food for the grassland predators: snakes, hawks, weasels, kit fox, and coyote.

Ants of the dry grasslands occupy the important soil-loosening and soil-turning niche which is filled by earthworms in moist soils. The mound-building harvester ants bring soil up from a depth of 6 to 8 feet. The mounds, surrounded by bare circles from which they have cleared the vegetation are a common feature of the New Mexico landscape.

Ground-nesting birds, like the quail, are typical of the grassland, where about half of the birds nest on the surface of the ground, and a third in weeds or shrubs. Some grassland birds even nest in holes in the ground, a niche unfilled in the forest. (Photos of Rodents and Ants by R.B.; quail by Penn. Game Commission)

![Desert biome photograph]

Desert biome is characterized by widely spaced drought-resistant evergreen plants which carry on photosynthesis all year, and by small annual plants that flower only in the spring rainy season and then die. The true deserts of California and Arizona, such as the Colorado Desert of California shown here, have as their dominant the creosote bush, *Larrea*, in association with the burro weed, *Franseria*. (Photo, R.B.)

The **smoke tree**, *right*, has no leaves and the pointed stems carry on photosynthesis. The plant at the *lower right* has no leaves; reduces evaporating surface by shedding portions of the older branches in the dry season. Shriveled stems are already evident in this photo taken in April. (R.B.)

Round-tailed ground squirrel, Citellus tereticaudus, of the Colorado Desert of California, comes out of its burrow and takes over the niche vacated when kangaroo rat retires for the day. (Photo by R.B.)

Nomadism is a human niche of great deserts like the Sahara of northern Africa. Since rain in such a desert does not fall everywhere at the same time or even in the same year, nomadic Africans wander from one favored place to another, grazing their few sheep or goats on vegetation that may receive effective rain only once in several years. The **camel**, which can go for 5 to 12 days without drinking water, is the vehicle of choice in the African desert. The traveller *above* is a French school inspector making his rounds of the travelling desert schools that serve the children of nomads in French West Africa. (Photo, courtesy UNESCO)

Jerboa, *below,* is the most drought-resisting mammal of the north African deserts. Like the kangaroo rat (see page 37) of American deserts, it is nocturnal, and has greatly enlarged hind legs for making great leaps across the sand. Both jumping and extreme swiftness in running is characteristic of desert rodents, which have to venture long distances from their burrows to gather enough food. The jerboa, like the kangaroo rat, can live for long periods on dry seeds, deriving its water from the water produced by its own metabolism. (Photo made in Bristol Zoo, England, by R. B.)

The tropical forest biome is characterized by evergreen trees of great height, their crowns forming an almost continuous canopy at the top, by two or more lower canopies formed by the crowns of lesser trees, and by several levels of woody shrubs. These divide the forest into a stratified series of somewhat separated communities supporting different assemblages of animals at the different levels. Great woody vines climb the trees, and a profusion of aerial roots hang from plants perched high overhead. At all seasons there are trees shedding leaves, flowering, or fruiting. (Panama forest. Photo by R.B.)

Evergreen seasonal tropical forest surrounds the Barro Colorado Island Laboratory in the Panama Canal Zone. It differs from equatorial rain forest in having a drier winter season in which many trees briefly drop their leaves, admitting light to support a more luxuriant ground stratum than in a rain forest. Deciduous trees do not all shed in winter season: on June 21 the tree at left was leafless; by July 7 it had put out a new set of foliage. (Panama, photo made from laboratory clearing looking toward the Canal. By R.B.)

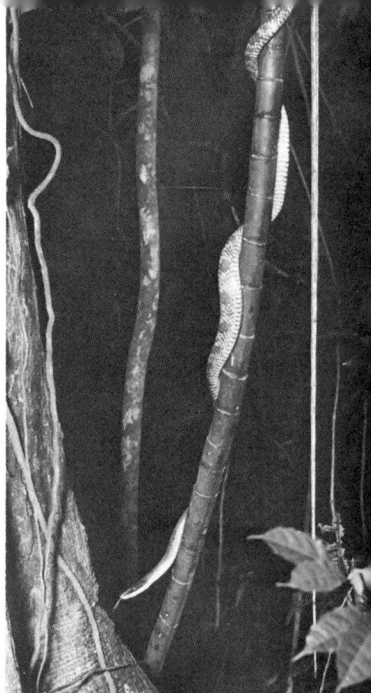

Sloth, in this case the 3-toed species, is one of the many kinds of animals unique to the South American tropical forest. Sluggish and dim-witted, it has survived, as have many primitive mammals in the tropics, by retreating to a small and specialized arboreal and nocturnal niche. It feeds almost exclusively on leaves of the Cecropia tree, and can sleep or munch leaves while hanging up-side down from tree branches, held securely by its huge curved claws. (Photo in Panama by R.B.)

Night monkey, or douricouli, is the only truly nocturnal monkey. Anyone who has heard its hooting favors its other name, the owl monkey. Its tail is not prehensile like that of other New World monkeys; but the grasping feet, large eyes and keen sense of smell enable this primate to get about in trees at night to find its fruit and insect food. (Photo of animal from Columbian forest by R.B.)

Tree snake is a common niche in the tropical forest, where a prehensile body makes climbing easy and trees are filled at many levels with birds' nests, and with small amphibians, reptiles, and mammals. (Panama. Photo by R.B.)

Bat clings to bark of tree in daytime; at night swoops actively about catching insects on the wing. In the year-around warmth of the tropical forest insect- or fruit-eating bats can feed at all seasons. (Photo,R.B.)

Protective resemblances are a noticeable feature of the tropical forest, where evolutionary change has had many millions of years to operate as compared with the recently evolved temperate forests. *Upper left,* a long-horned **beetle** has markings that make it resemble the tree bark on which it rests. *Upper right,* a 12-inch **walking-stick** (belongs with grasshoppers to insect order Orthoptera) resembles the climbing vines among which it was found. *Lower left,* a brown **mantis** that resembles the dead leaves among which it lives. It has numerous mantis relatives that are green and resemble the green leaves of their habitat. *Lower right,* a harmless **moth** mimics a bee. (Photos made in Panama forest by R. B.)

Lizard is one of tremendous variety of arboreal reptiles that occupy the many niches of a forest in which most plants are woody and tree-like in size and there is no period too cold for activity by cold-blooded animals. Body 3 in., tail 9 in. long. (Panama. Photo by R.B.)

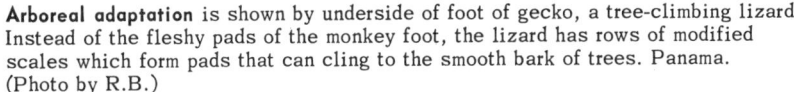

Arboreal adaptation is shown by underside of foot of gecko, a tree-climbing lizard. Instead of the fleshy pads of the monkey foot, the lizard has rows of modified scales which form pads that can cling to the smooth bark of trees. Panama. (Photo by R.B.)

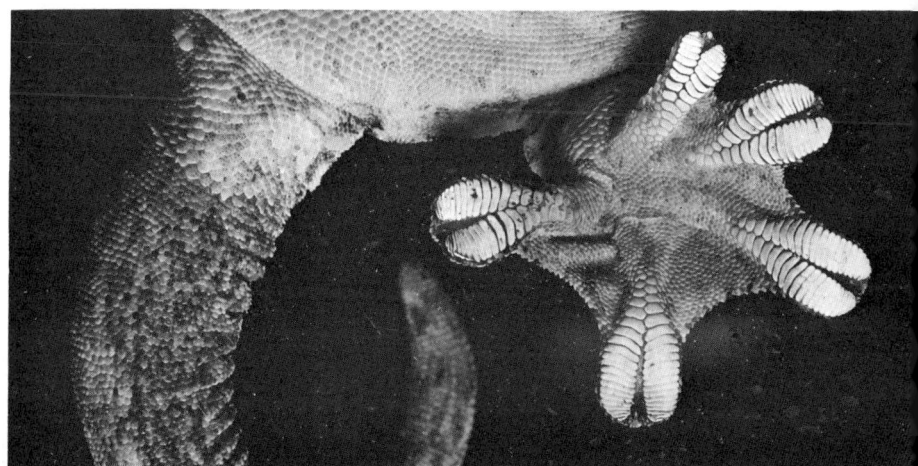

THE MARINE BIOME

The considerations on which terrestrial biomes are based do
not apply well in the oceans, where plants do not exert the controlling influence that they do on land. The seas are all intercon-
nected, temperature variations are not of great range, and moisture is a limiting factor only on tidal shores. Though the oceans
cover two and a half times as much of the earth's surface as do
all the land masses, and support a greater bulk of living organisms,
ecologists find it best to consider the oceans as a single marine
biome.

The oceans can be divided into two major divisions. The benthic
division includes all of the bottom-living communities of the ocean
floor from the shore down to the deepest abyss. The pelagic division includes the whole mass of water in which plants and animals
float or swim. The benthic division can be further subdivided into
a littoral benthic zone extending from high tide mark down to the
edge of the continental shelf and a deep-sea benthic zone extending
from the edge of the continental shelf to all the depths of the ocean.
The pelagic division is comprised of a neritic pelagic zone from
the shore to the edge of the continental shelf and an oceanic pelagic
zone beyond the edge. The dividing line between the littoral and the
deep sea usually comes at about 200 meters (600 feet) and this is
also roughly the maximum depth to which light penetrates.

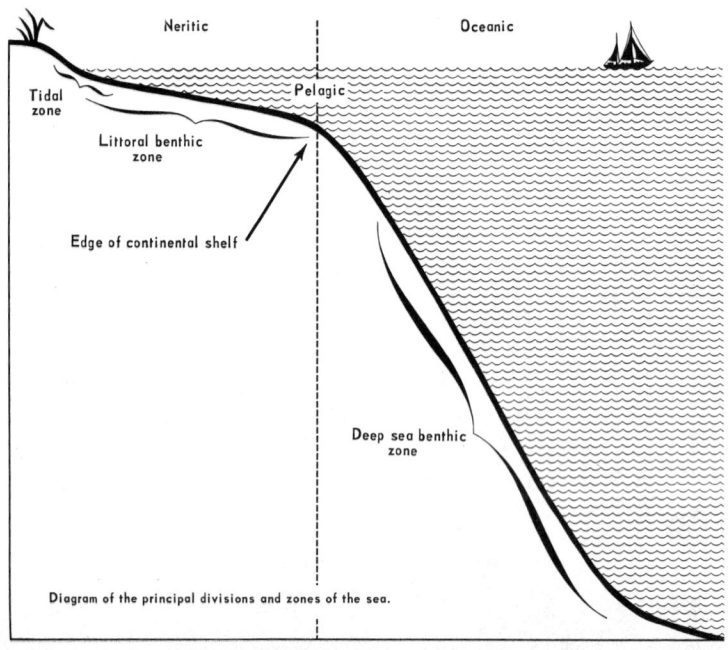

Diagram of the principal divisions and zones of the sea.

The sea shore is the meeting place of land and sea. Twice daily the tide covers
and then uncovers the plants and animals of the intertidal area between high and
low tide marks. The tidal community must adapt to abrupt changes from land to
marine climates. On a **protected muddy shore** (as *above*, on the coast of South Carolina) the ebb and flow of the tide keeps stirring up the surface and to maintain
position animals must burrow into the more stable mud below. Clams, worms, and
snails remain in their water-filled burrows when the tide is out; only the fiddler
crabs come out to scavenge organic material.

On an **exposed rocky shore** (near Ensenada on the west coast of Lower California)
the pounding surf smashes unattached animals against the rocks or casts them up
on the beach. Where the surf is strong there are few algae, and the only successful animals are permanently attached barnacles and mussels or those capable of
holding on securely like snails, limpets or chitons. (Photos by R.B.)

Temperate rocky shore at Pacific Grove, California. The gulls and cormorants are land animals that feed entirely from the sea. Their white droppings, deposited on the rocks, are washed back into the waters from which the organic material came. (Photo by R.B.)

ZONATION ON A ROCKY TEMPERATE SHORE

The difference in climate from yard to yard of a steeply sloping shore may be that of habitats that are thousands of feet apart in the climatic zones of mountains or many hundreds of miles apart in the temperature and moisture gradients that occur on land.

The orderly zonation of plants and animals is so similar for rocky coasts of temperate climate the world over, the organisms so alike in appearance and in niche, that it often takes an expert to notice that the species or even the genera are different. The rocky coasts of South Africa, of the east and west coasts of North America, and especially those of England, have been studied and their similarities pointed out by the Stephensons. In the photo above we can see the dark uppermost zone, the <u>Littorina zone</u>, of the tidal area. The gray lichens, the small snails and the small barnacles in it are wetted by spray and by the highest tides. Below it is the large whitish <u>balanoid zone</u> covered and uncovered by most tides. At the water's surface we can see parts of the large brown algae of the <u>subtidal algal zone</u> below low water mark.

SNAILS

🐚 Littorina neritoides

🐚 L. rudis

🐚 L. obtusata

🐚 L. littorea

BARNACLES

🐚 Chthamalus stellatus

🐚 Balanus balanoides

🐚 B. perforatus

ALGAE

a. Pelvetia canaliculata
b. Fucus spiralis
c. Ascophyllum nodosum
d. Fucus serratus
e. Laminaria digitata

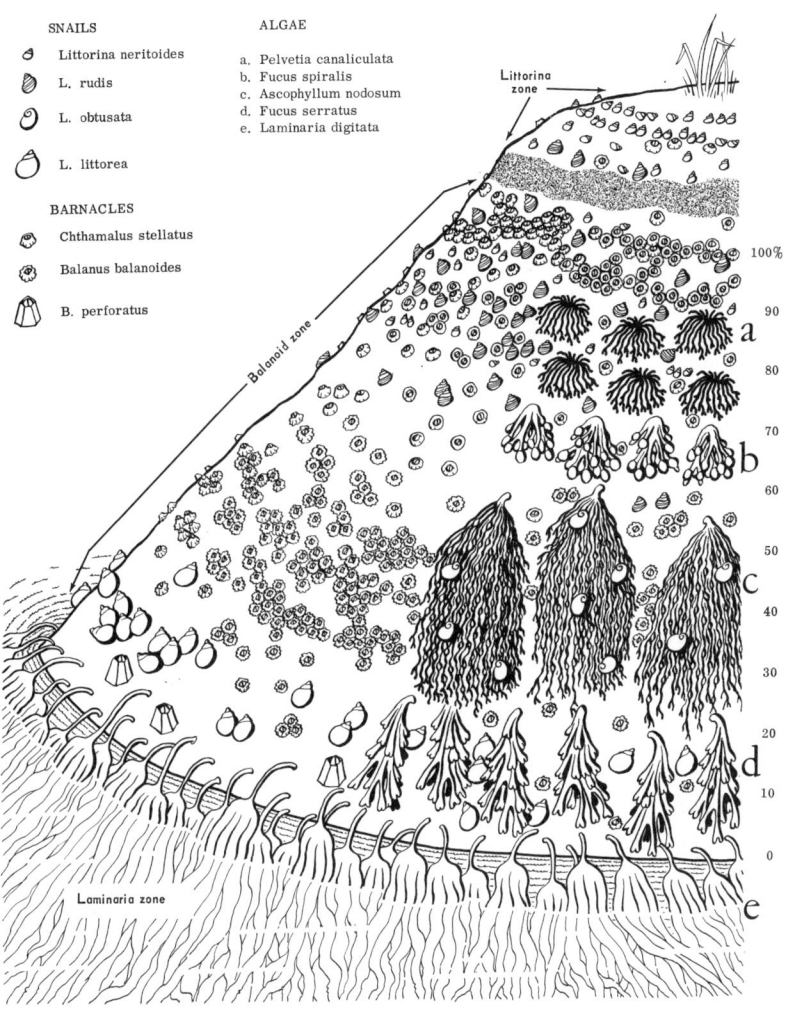

Zonation of periwinkle snails, barnacles, and algae on a rocky shore. This situation is that of the southwest coast of England, but the same genera of periwinkles and of barnacles occur on American coasts and the zonation is comparable. Left half of diagram, rocks with exposure to surf too great to support algae; right-hand half is of less exposed rocks, where algae obscure most of the animals. Numbers at the right-hand edge are percentages of time the zone is exposed to air during the whole of a year and are based on Colman's data for the Wembury coast in the south of England. Zones and their names are based on Stephenson.

Algae of the balanoid zone are of various species, which differ with geography and with exposure to surf and other factors; but order of zonation is consistent. Wave-beaten shores are devoid of algae, but wherever rocks are partially protected there grow the most surf-tolerant brown algae, like *Fucus vesiculosis*.

Animals of the balanoid zone shown here are an acorn barnacle, *Balanus balanoides;* a carnivorous snail, *Nucella lapillus,* that feeds on barnacles and mussels; and a limpet, *Patella vulgata,* that uses its radula to scrape algae and minute animals from the rocks. (Photos made at Trevone, England, by R.B.)

Animals of sandy shores are mostly worms, clams, snails, and crustaceans that maintain their position in the sand by burrowing below the shifting surface. The **tube anemone**, *Cerianthus*, maintains itself just above the surface by encasing the elongated body in a parchment-like tube of secreted material. One California species has a tube known to go six feet deep. (Photo by R.B. in Naples Aquarium, Italy.)

Clams of many kinds live below the surface of sandy beaches in the damp zone between tidemarks, but they remain in their burrows when the tide is out and are seldom seen unless they are dug out. Coquinas or wedgeshells, *Donax variabilis*, are about 1 in. or less long. (Beaufort, N.C. Photo by R.B.)

Crab of sandy beach is one of few animals of an exposed sandy beach that comes out of its burrow at low tide. The "ghost crab," *Ocypode*, occurs on our East Coast from New Jersey southward.(Panama. Photo by Mildred Buchsbaum.)

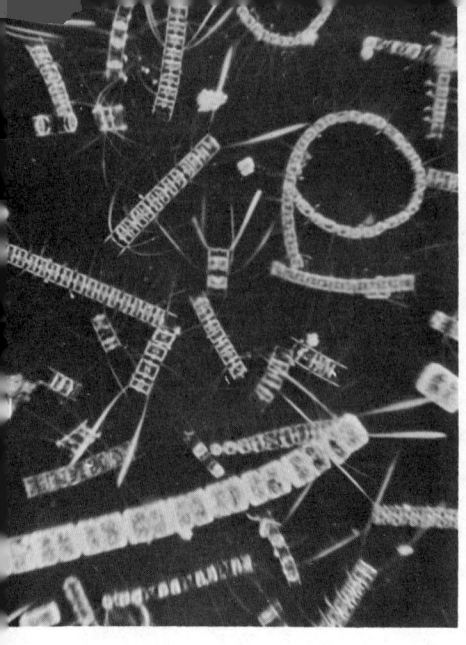

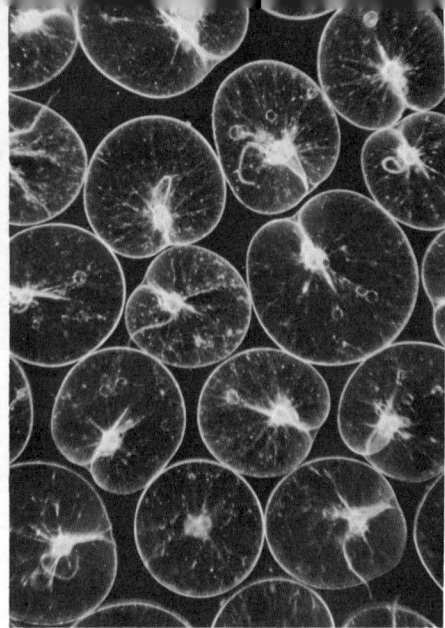

Plant plankton serves as the major basis of life in the sea. This sample is mainly species of *Chaetoceros*. (Photos by D. P. Wilson.)

Key-industry animals of the ocean are the copepods, of which the most common are *Calanus*, the larger crustaceans shown here, which are about 1/6 inch in length.

Protozoans constitute a large part of the animal plankton; here is *Noctiluca scintillans*, a luminescent dinoflagellate.

Other animal plankton include, as shown here, in order of increasing size: nauplius larvas of crustaceans, copepods, older larvas of crabs, and a long slender arrow worm.

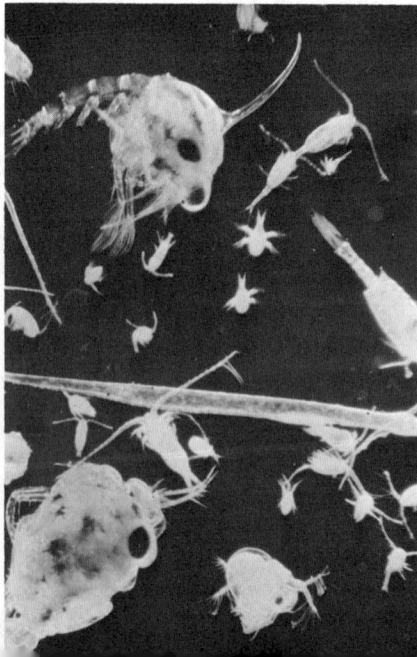

The pelagic community of the ocean is illustrated here by scup (porgies) photographed near the bottom at 480 feet in the Gulf of Mexico. Pelagic animals may feed on the benthic community but most of them feed on floating or freely swimming pelagic animals of intermediate depths. (Photo Woods Hole Oceanographic Institute)

The benthic community, or bottom-living plants and animals of the sea, are most numerous close to shores, where the waters are enriched by organic matter from the land. In shallow waters the littoral benthic community receives enough light to support many algae. At 240 feet, in the deeper waters of the continental shelf (Georges Bank, off Massachusetts), the **sand dollars**, *below left*, feed on diatoms and other organic materials that settle down from above. In the deeper waters of the ocean, the deep-sea benthic community is sparser, but no haul fails to bring up some organisms. *Below right*, a large sea spider (pycnogonid) and several small serpent stars (echinoderms) on the bottom at 6000 feet in the waters off Cape Cod. (Photo by D.M.Owen, Woods Hole Oceanographic Institute.)

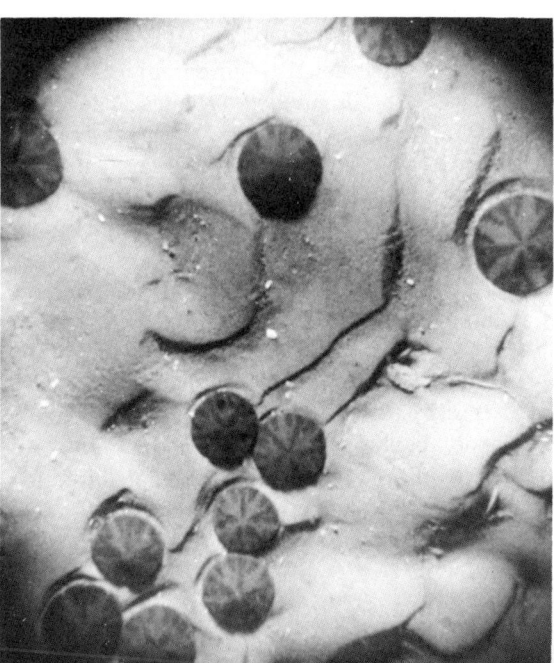

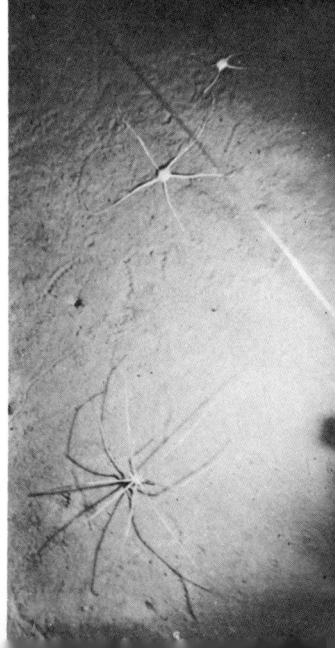

Chapter 9

CLIMATIC GRADIENTS IN PLANTS AND ANIMALS

THE DIRECTNESS with which plants meet the impact of the physical environment has already been referred to. Behavioral adjustments play almost no role, and adaptation is almost entirely structural. Limitations of inherited structure that prevent palms or bananas from surviving Pennsylvania winters are best known to us. In southern California peach varieties from the colder Midwest cannot grow, because the winters are too mild and do not provide sufficient cold to break the dormant period. Adaptations to limited water supply that reduce the size of leaves or that eliminate leaves altogether, were described in the discussion of the desert biome. One other major climatic gradient needs to be mentioned, that of the length of day required for flowering. In general, plants that have originated in tropical latitudes will flower only when the days are short and will fail to flower when the lighted part of a day-night cycle exceeds a certain number of hours. Soybeans flower when they receive 8-hour days and 16-hour nights, not because the day is short but because the dark period is sufficiently long. If the dark period is interrupted by a few minutes of light from an incandescent bulb, the soybeans will not flower. Such plants are called short-day plants. Plants that have originated in northern latitudes (north of 60°) usually flower only when the lighted part of a day exceeds a certain number of hours. The length of the dark period is of no importance, and they can flower with continuous light. Such plants are called long-day plants, and we can see that they would fit the requirements of the tundra, where there is continuous light in summer. A third group of plants seems not to be influenced at all in flowering by length of day. And in temperate latitudes there are these indeterminate types, short-day plants, and long-day plants.

Animals have many behavioral devices for circumventing the physical conditions of their climate. They can shorten their hours of daylight by climbing under bark or into a darkened log or a cave. They huddle together for warmth, burrow underground, or build warm shelters when the temperature drops. They can move about in search of water, while plants die of drought only a few yards from a stream that flows through arid land. There are, however, certain limits to behavioral adaptation, and animals have also become structurally evolved to meet the physical conditions of the various climates in which they live.

As in plants, many structural adaptations to climate have become fixed by heredity and develop even when climate is changed experimentally. A polar bear born and reared in a zoo in Pennsylvania will still have heavy white fur and furred paws, though his coat may never be quite as thick as that of a bear that lives through polar winters.

On the other hand, the larger heart size/body size ratio found at higher latitudes may be a physiological adjustment, and reversible with change in climate. Chickens that are experimentally reared at low temperatures develop larger hearts in proportion to body size than do chickens reared at higher temperatures. Chickens reared at 43°F. during their early months also have stockier and heavier bodies and shorter legs and tails than do the same breed reared during all of their early months at 73°F.

Many plants, also, have different growth forms when reared in different climates. To determine the relative importance of heredity and of physiological adjustment in plant or animal structural adaptation usually requires carefully controlled experimentation. Nevertheless, structural differences do exist in different climates, and without saying much of exactly what internal mechanisms produce them, we shall here briefly consider a few of the climatic gradients which have been noted, especially those in animals, of which little was said in describing the biomes.

NUMBER OF SPECIES AND OF INDIVIDUALS

Excepting regions of aridity, the increasing warmth and moisture and light from poles to equator favors a marked increase in the numbers of plant and animal species, as we have seen. The increasingly favorable climate supports not only a greater variety but also a greater total bulk of plants, and so of animals. But the numbers of individuals in each species is another matter. The rigorous requirements of climate in tundra or in the grassland, or even in the temperate forest as compared with the tropical forest, are met by a lesser variety of plants and this in turn leads to the production of tremendous numbers of the same kind of plant: hundreds of miles of similar lichens, or of a few kinds of grasses, or of spruces and firs, or of oaks mixed with a few other trees. The abundance of one kind of plant food promotes the multiplication of whatever animals are adapted to feed on these plants, and so a Kansas grassland (or the even more homogeneous wheat crops of man) will support millions upon millions of the same species of grasshopper. In the tropical forest, where innumerable kinds of plants grow singly instead of in large stands, any one species of animal has to move about a great deal to feed on its plant food, and the multiplication of an animal species is limited by the relative scarcity of any particular kind of food. The number of different species in the tropics is tremendous, but the numbers of individuals in a species is relatively small.

In the oceans there is also a gradient of increasing numbers of species of invertebrates from the poles towards the equator. But the increase is concentrated among the invertebrates of rocky shores. On the level sea bottom, which occupies more than half the global area, and has relatively constant conditions, there is no increase in number of invertebrate species from the arctic sea bottom to the tropical deep sea bottom. One group of mollusks, has, according to Thorson, 84 known species in Greenland, 200 known species around southern England, and about 700 species from the Persian Gulf. But if we select, from among this larger group which includes animals of tidal shores, a certain sub-group of bottom-living mollusks, then the number of species is practically the same from the arctic to the equator. This applies also to the total weight of bottom-living forms, which is about the same on the level bottom off Greenland as in the Persian Gulf.

The bottom fauna of the continental slope does decrease in diversity and in total bulk as the dredge is lowered farther and farther down the slope. This is not so much a matter of depth as of increasing distance from the more plankton-rich coastal waters. Deep waters close to shore will have more bottom forms than will waters of the same depth that are at a greater distance from shore. Where currents bring plant nutrients farther out to sea, the extra growth of plankton at the surface supports a richer bottom fauna.

SIZE OF ANIMALS, ETC. IN RELATION TO TEMPERATURE

Though there are exceptions, the sizes of warm-blooded animals, the birds and mammals, are greater in colder regions and become increasingly smaller towards the equator. The statement of this tendency has been named Bergman's rule. The correlation between low temperature and large body size in warm-blooded forms is thought to be an adaptation to conservation of body heat by a reduction of body surface in proportion to body volume. In warm climates small body size gives greater body surface in relation to body volume and presumably facilitates radiation of heat. The emperor penguin that lives and breeds on the antarctic shelf at -18°C. to -62°C. (0° to -80°F.) averages 70.5 pounds in weight. Its relative, the king penguin, lives farther north in the southern hemisphere, where temperature is usually above 32°F. and only rarely below 0°F. It averages only 44 pounds in weight. Such size differences can be shown for all kinds of warm-blooded forms which have relatives in the various biomes.

Bergman's rule is inverted for cold-blooded land forms, which are generally largest in the tropics. Similar rules hold for animals at various altitudes on tropical mountains that have the same climate the year around. In the high mountains of New Guinea the biggest bird specimens are found at the highest, coolest levels. But the biggest grasshoppers of any given species are found nearest the base and the smallest specimens near the summit. Such differences cannot be established so easily in temperate regions,

A light outside a tropical laboratory in Panama, *left*, may in two evening hours attract about 700 insects, representing as many as 250 different species, *right*. In a temperate forest a comparable collection might yield only two or three dozen different species. (Barro Colorado Island, Panama. Photo by R.B.)

Size of ears in foxes tends to be small in the arctic, intermediate in temperate zone foxes like the red foxes, *left,* from Pennsylvania, and large in desert foxes like our southwestern kit fox and the fennec or North **African desert fox,** *right.* (Photos: red fox, Penn. Game Com.; desert fox, in zoo, Paris, by R.B.)

where animals tend to migrate up and down mountains as the seasons change.

Size of marine animals varies with temperature also. Marine copepods are larger at low temperatures, smaller at high temperatures. Starfishes and jellyfishes are larger in cold waters than in warm ones. These size differences are not necessarily directly related to temperature, or even to the same factors in all cases. Low temperature tends to delay reproduction and so produce large animals. And warm and cold waters also differ in salt content and in viscosity. Warm waters are therefore less buoyant than cold, and for small animals size may be importantly related to buoyancy of the water. This usually is given as the explanation for the increased size in warm waters of the protuberances that assist animals in flotation.

Size of body extremities in closely related warm-blooded animals decreases from south to north. This is the reverse of the trend in total body size, but is related to the same temperature factor, and is called Allen's rule. The hares of the genus Lepus (commonly known as jackrabbits) are represented by short-eared species in the arctic and by progressively longer-eared species from the northern to the southern United States.

The emperor penguin has feet, wings, and bill of about the same absolute size as those of the much smaller king penguin, so that the extremities are proportionally smaller in the penguin that lives at lower temperatures. This again seems to be a matter of heat conservation in cold climate by reduction of body surface in relation to body volume. The larger ears of the desert fox presumably increase the radiation of heat.

Number of vertebras in closely related fish is one of many quantitative gradients that are correlated with latitude. Fishes that live at low temperatures tend to have more vertebras than do fishes of warmer waters. The correlation is regular enough to be referred to as Jordan's rule, but it is complicated by the fact that in some fishes number of vertebras is related to total body size, and that size may vary with factors other than temperature.

Unbalance in animal numbers is so frequently caused by man in modern times, that we sometimes overlook the fact that animal plagues probably occurred long before man appeared. Any change in weather which tips the balance in favor of some one plant or some one animal will also result in an increase in the other animals of the food web. Small herbivores, especially mice or lemmings or rabbits, increase tremendously in numbers when their plant food increases. The physical factors which increase the plant food are in many cases obscure, but they do seem to occur at regular intervals, so that "lemming years" or "arctic hare" years can be more or less predicted in advance. When lemmings increase in numbers so do all the predators which feed on them. And since many of the predators are valuable fur-bearers, the records of fur-trappers from year to year have served as a convenient source of data for studying the cyclic increases that occur in northern regions. Tremendous increases in numbers of rodents, of grasshoppers, of toads, and of other animals occur in temperate regions but they are not so clearly periodic nor so well documented.

The point to be made here is that great variations in animal numbers tend to occur in regions of climatic extremes such as arctic regions, or our great plains, or in deserts, where weather varies greatly from year to year and a year of drought or one of very heavy rainfall can set off a plague. Sharp fluctuations in animal numbers are practically unknown in the wet tropics. Plagues occur most readily where the number of different kinds of niches is limited. The natural regulation of animal numbers is accomplished by a system of natural checks and balances which many ecologists are at present trying to analyze. We should wish them success for there will soon come a time when man will have to decide whether to respect these natural mechanisms developed over many millions of years or whether he can go on pretending that he is forever exempt from the rules of the ecological system of which he is clearly a part.

To sum up, then, the ecological viewpoint as applied to man's practical problems is the conservative view of man's relation to his total environment. It holds that an environmental setting as complicated as a planet inhabited by more than a million and a half species of plants and animals, all of them living together in a more or less balanced equilibrium in which they constantly use and re-use the same molecules of the water, land, and air, cannot be improved by aimless and uninformed tinkering. All changes in a complex mechanism involve some risk and should be undertaken only after careful study of all of the facts available. Changes should be made on a small scale first, so as to provide a test, before they are widely applied. When information is incomplete, changes should stay close to the natural processes which have in their favor the indisputable evidence of having supported life for a very long time.

BIBLIOGRAPHY

TECHNICAL JOURNALS

Ecology. Duke University Press, Durham, N. C.
Ecological Monographs. Duke University Press, Durham, N. C.
Journal of Ecology. Cambridge University Press, N. Y.

POPULAR JOURNALS

Endeavor. London.
Natural History. New York.
The New Scientist. London.
Science News Letter. Washington.
Scientific American. New York.

TEXTBOOKS

Allee, W. C., A. E. Emerson, O. Park, T. Park, and K. P. Schmidt
Principles of Animal Ecology. Saunders Co., Philadelphia. 1949.

Benton, A. H. and W. E. Werner, Jr. Principles of Field Biology and
Ecology. McGraw-Hill Co., N. Y. 1958.

Clarke, G. L. Elements of Ecology. John Wiley and Sons, N. Y. 1954.

Clements, F. E. and V. C. Shelford. Bio-ecology. John Wiley, N. Y. 1939.

Daubenmire, R. F. Plants and Environment. John Wiley, N. Y. 1959.

Elton, C. Animal Ecology. Macmillan Co., N. Y. 1927.

"   The Ecology of Animals. John Wiley, N. Y. 1950.

Macfadyen, A. Animal Ecology. Pitman, London. 1957.

Odum, E. P. Fundamentals of Ecology. Saunders, Philadelphia. 1959.

Oosting, H. J. Plant Communities. Freeman, San Francisco. 1956.

Weaver, J. E. and F. E. Clements. Plant Ecology. McGraw-Hill, N. Y.
1929.

Woodbury, A. M. General Ecology. Blakiston, Philadelphia. 1954.

SPECIAL ASPECTS OF ECOLOGY

LAND

Aubert de la Rue, E. , F. Bourliere, and J. P. Harroy. The Tropics. Knopf, N. Y. 1957.

Buxton, P. A. Animal Life in Deserts. E. Arnold, London. 1923.

Collins, W. B. The Perpetual Forest. Lippincott, Philadelphia. 1959.

Folsom, F. Exploring American Caves. Crown, N. Y. 1956.

Jaeger, E. C. The North American Desert. Stanford, Calif. 1957.

Richards, P. W. The Tropical Rain Forest: an ecological study. Cambridge Univ. Press, London. 1952.

Sears, P. B. Deserts on the March. Univ. of Oklahoma Press. 1937.

Weaver, J. E. and Albertson, F. W. Grasslands of the Great Plains. Johnsen Pub. Co. , Lincoln, Nebraska. 1956.

FRESH WATER

Coker, R. E. Streams, Lakes, Ponds. Univ. of North Carolina Press, Chapel Hill. 1954.

Pennak, R. W. Fresh-water Invertebrates of the United States. Ronald Press, N. Y. 1953.

Welch, P. S. Limnology. McGraw-Hill, N. Y. 1935.

MARINE

Buchsbaum, R. Life in the Sea. Univ. of Oregon, Eugene. 1955.

Coker, R. E. This Great and Wide Sea. Univ. of N. Carolina Press. 1947.

Colman, J. S. The Sea and Its Mysteries. Norton, N. Y. 1950.

Hardy, A. C. The Open Sea. The World of Plankton. Collins, London. 1956.

"   The Open Sea, II. Fish and Fisheries. Collins, London. 1959.

Hedgpeth, J. W., Ed. Treatise on Marine Ecology and Paleoecology. Vol. I. Ecology. Geol. Soc. Amer. Memoir 67, N. Y. 1957.

Moore, H. B. Marine Ecology. John Wiley, N.Y. 1958.

Ricketts, E. F. and Calvin, J. Between Pacific Tides. 3rd. ed. rev. by J. W. Hedgpeth. Stanford Univ. Press. 1952.

Sverdrup, H. U. , M. W. Johnson, and R. H. Fleming. The Oceans, their Physics, Chemistry and General Biology. Prentice-Hall, N. Y. 1946.

Wilson, D. P. Life of the Shore and Shallow Sea. Nicholson and Watson, London. 1951.

Yonge, C. M. The Sea Shore. Collins, London. 1949.

DISTRIBUTION

Andrewartha, H. G. and L. C. Birch. The Distribution and Abundance of Animals. University of Chicago Press. 1954.

Cain, S. A. Foundations of Plant Geography. Harper, N. Y. 1944.

Dansereau, P. Biogeography: An Ecological Perspective. Ronald Press, N. Y. 1957.

Hesse, R. , Allee, W. C. and K. P. Schmidt. Ecological Animal Geography. John Wiley and Sons, N. Y. 1937.

Newbigin, M. I. Plant and Animal Geography. Dutton, N. Y. 1948.

BEHAVIOR OF ANIMALS

Allee, W. C. Cooperation among Animals. Schuman, N. Y. 1951.

Carthy, J. D. An Introduction to the Behaviour of Invertebrates. Allen and Unwin, London. 1958.

Frisch, Karl von Bees: Their Vision, Chemical Senses, and Language. Cornell Univ. Press, Ithaca. 1950.

Scott, J. P. Animal Behavior. Univ. of Chicago Press. 1958.

Tinbergen, N. Social Behaviour in Animals. Wiley, N. Y. 1953.

Wheeler, W. M. Social Life among the Insects. Harcourt, Brace, N. Y. 1923.

CONSERVATION

Bennett, H. H. Elements of Soil Conservation. McGraw-Hill, N. Y. 1947.

Dasmann, Raymond F. Environmental Conservation. Wiley, N. Y. 1959.

Huberty, M. R. and W. L. Flock. Natural Resources. McGraw-Hill, 1959.

Leopold, A. Game Management. Scribner's Sons. , N. Y. 1933.

Milne, Lorus J. and Margery J. The Balance of Nature. Knopf, N. Y. 1960.

Peterson, Elmer. Big Dam Foolishness: The problem of modern flood control and water storage. Devin-Adair, N. Y. 1954.

Thomas, W. L. Jr., Ed. Man's Role in Changing the Face of the Earth. U. of Chicago Press, 1956.

MISCELLANEOUS

The Australian Environment. Com. Sci. and Industrial Res. Org. Melbourne. 1950.

Bews, J. W. Human Ecology. Oxford Univ. Press, London. 1935.

Buchsbaum, R. Animals Without Backbones. Univ. of Chicago Press. 1949.

Cott, H. B. Adaptive Coloration in Animals. Oxford Univ. Press, N. Y. 1940.

Dice, L. R. Natural Communities. Univ. of Michigan Press, 1952.

Elton, C. Voles, Mice and Lemmings. Problems in Population Dynamics. Clarendon Press, Oxford. 1942.

"   The Ecology of Invasions by Animals and Plants. Methuen, London. 1958.

Hanson, H. C. and E. D. Churchill. The Plant Community. Reinhold, N. Y. 1961.

Henderson, L. J. The Fitness of the Environment. Macmillan, N. Y. 1913.

Lack, D. Natural Regulation of Animal Numbers. Oxford, London. 1954.

Malthus, T., J. Huxley, F. Osborn. On Population: three essays. Mentor, N. Y. 1960.

Milne, Lorus J. and Margery J. The World of Night. Viking, N. Y. 1956.

Reid, L. Earth's Company. John Murray, London. 1958.

Sears, P. B. Life and Environment. Columbia, N. Y. 1939.

Shelford, V. C. Animal Communities in Temperate America. Univ. of Chicago Press. 1913.

U. S. Dept. Agriculture Yearbooks: e. g., 1938 Soils and Man; 1940 Farmers in Changing World; 1941 Climate and Man; 1948 Grass; 1949 Trees; 1952 Insects; 1956 Animal Diseases; 1955 Water; 1958 Land. Washington, D. C.

Watts, May T. Reading the Landscape. Macmillan, N. Y. 1957.

INDEX

Numbers in *italics* refer to pages which bear illustrations of the item listed.

A

Adaptations to
  climate, 178, 179
  physical factors, 21
Aestivation, 93
Aggregations, *52-54*
Air currents affecting climate, *144*
Algae,
  brown, 26, *173, 174*
  and *Convoluta, 45*
  and light, 32, 170
  in protozoan succession, 101, *102*
  sea shore zonation, *173, 174*
  and tissue culture, 81, *82*
Allen's rule, 182
Ameba in protozoan succession, 101, *102*
Anemone, tube, *175*
Annual rings, 129
Antarctic, *154, 155*
  penguins, *24, 155*, 181, 182
  seals, *155*
  weather station, *155*
Ant,
  and aphids, *51*
  eater, 46, *68*
  in grassland, *163*
  societies, *56*
Aphid, *43*
  eating of, 66
  tending by ants, *51*
Arboreal adaptation, *169*
Arctic,
  bear, *see* polar bear
  hare years, 183
  seasonal changes, 90
Arid, *see desert*
Arrow worm, 20, *176*
Athletes' foot, 45

B

Bacteria,
  in protozoan succession, 100, *102*
Badlands, 12
Balance,
  in animal numbers, 183
Balanoid zone, 172, *173, 174*
Bare bottom, 108, *111*
Bare area, 127
Barnacles, 172, *173, 174*
  in Salton Sea, *140*
Barn owl, *3*

Barn swallows, *53*
Barriers, 136-140
Bat, 34, *167*
  sonar, 85
Bear, black, *156*
  in sleep, *30*
Bear Island, 74, *75*
Bear, polar, *see* Polar bear
Beaver, *159*
Beech and maple forest, 118, *119, 121, 123*
  121, 123
  as temperate forest biome, *157*
Beetle, dung, *68*
Beetle,
  ladybird, 18
  tropical, *168*
Benthic community, *177*
Benthic division, 170
Bergman's rule, 181
Bering Sea land bridge, 142
Biogeography, 133
Biome, 141-163
  coniferous forest, 148
  deciduous forest, 148
  diagram of world, 150
  desert, 148, *164, 165*
  grassland, 148, *162, 163*
  land, 147
  marine, 170-177, *171-177*
  rain forest, 150, *166*
  temperate forest, 148, *144, 119, 123*
  tundra, *148*
Biotic environment, 39-57
  factors, as barriers, 140
Bird,
  ground-nesting, 163
  migration, 94
Bison, *162*
Blowout, *123*
Bobwhite, *163*
Burro weed, *164*
Burrowing lizard, *36*

C

Cactus, *15*
  water adaptations of, 38
Caecilian, *see* Cecilian
Camel, *165*
  dispersion of, 133
Carbon cycle, 79
Caribou, lichens for, *153*
Cave animals, 33, *35*
Carnivore, 63, 64, *65*
  top, 78
Cecilian, *25*

Cereus, night-blooming, *87*
Chaparral, 149
Chara, 108, *111*
Chickens, peck-order in, 55, *57*
China, 8
Chlorella, mutualism with, 81, *82*
Clams, sandy shore, *175*
Climate,
  as barrier, 139
  gradients, 178-183
Climax, 100, 122
Climax stage,
  in dune succession, 118, *119, 121, 123*
  in pond succession, *111*, 112
  in protozoan succession, 101, *102*
Cold blooded animals, 28, 92
  Bergman's rule for, 181
  numbers of, 180
Colpoda, 101, *102*
Commensalism,
  animal, 47, *48, 49*
  plant, 40
Community, 58-82
  human, 61, 126, *127*
  log, 59
  oak, 58
  naming of, 61
  pond, 59
Concealing coloration 33
Concealing structure, *168*
Coniferous forest biome, 148, *156*
Conservation, soil, 8
Consumers, food, 62
Contour planting, *11*
Convergent evolution, 70
Convoluta, *45*, 67
Cooperation, 52
Copepod, *176*, 182
Cores, ocean, *131*
Cormorants, *172*
Cottonwood, 115, *119, 120*
Crab, sandy beach, *175*
Crayfish, cave, *35*
Creosote bush, *164*
Cricket, *65*
  thermometer, *29*
Crop parasites, 17
Cycle, carbon, 79
Cycle, of animal numbers, 183
Cypress, *40*

D
Dam, 2, 16
Day length, 178
Day-night, 84
  in desert, 87
  in equatorial regions, 88

Day-night,
  in oak-hickory wood, 85
  in ocean, 89
  in plants, 33, 86
  in polar area, 88
Deciduous forest biome, 148
Decomposer, niche of, 64
Deep sea zone, 170, *177*
Deer, white-tailed, 20, *158*
Desert biome, 148, *164, 165*
  day-night, 87
  plants, *37, 38, 164*
  seasonal changes in, 91
  shadscale, 151
  water, 36
Desert fox, *182*
Diatoms, *176*
Disclimax, 122, 125
Discontinuous distribution, 130
Dispersion, 133, 134, *140*
Distribution
  barriers, 136-140
  accidental, 135, *140*
  discontinuous, 130
  of plants and animals, 128-140
  of polar bear, 128
  rate of, 134
  of whales, 129
Dolphin, *24*
Dominant.
  plant, 60
  species, 39
Douricouli, *167*
Drift ice, 128
Drift line, 113
Drought, *10, 164*
  see also, Desert
Dune succession, 113, *119, 120, 121, 123*
Dung beetle, *68*
Dusk, 86
Dust storm, *11*, 13

E
Echiuroid worm, *47*
Ecological succession, 98-127
Ecological viewpoint, 1-20, 9
Ecologist, role of, 5
Ecology, definition of. 3
Ecosystem, 21, 58, 61
Ecotone, 112
  forest-tundra, *153*
Eggs, frog, *44*
Elm tree, *40*
Emerging vegetation, 109, *111, 112*
Energy, struggle for, 39
Energy, total, 74, 78
Environment, biotic, 39-57

Environment, biotic, 39-57
Epiphyte, *40, 41*
Erosion, 12, 16
  and overgrazing, *14,* 15
  wind, *11*
Evergreen tropical forest, *166*
Evolution, convergent, 70
Evolution and ecology, 21

F

Fallen tree succession, 103, 104
Farm, abandoned, *10*
Fish,
  cave, *35*
  hatcheries, 19
  number of vertebras, 183
Fishing, 18
Flagellates, 101, *102*
Flatworm,
  cave, *35*
  Convoluta, 45
Floatation, 23, 182
Flood, 2, 12, *14,* 13
  control, 16
  damage, 5
Food chain, 70, 71
Food consumers, 62
Food producers, 62
Food pyramid, *front cover, 76*
Food web, 72
  Bear Island, *75*
  herring, *77*
Foraminifera, 130, *131*
Fore dune, 114, *119, 120*
Forest, 148, *156*
  biomes, 148
  beech-maple, 118, *119, 121, 123*
  coniferous, 148, *156*
  deciduous biome, 148
  distribution in U.S., *144*
  of Ohio, 9
  pine, 116, *119, 121*
  temperate, 91, 147
  tropical, *166*
Fox,
  arctic, 182
  desert, 182
  diet of, *42*
  red, *182*
Frog,
  eggs, *44*
  leopard, *65*
  tree, *44*
  wood, *157*
Fucus, 174
Fungus, 45

G

Gannets, *54*
Garter snake, *65*
Gecko, foot of *169*
Geographic barriers, 146
Giraffe, 146
Glaciation, 129
Glass sponge and crab, 47, *48*
Grasslands, *10*
  biome, 148, *162, 163*
  distribution in U.S., *144*
Grazing, 15
  in China, 9
Ground squirrel, *164*
Growth rings, 129
Gulls, *172*

H

Habitat, 22
Hatcheries, fish, 19
Hawaii, 136
Hawk, *65*
Heart size/body size, 179
Heat, 27; *see also* Temperature
Herbivore, 42, 43
Herring food web, *77*
Hermit crab, 49
Herring food web, 20, *77*
Hibernation, 30
  in polar regions, 91
Homothermalism, 29, 96
Human communities, 83
Human succession, 126, *127*
Humidity, 29
Hunting, 18
Hypotrichs, 101, *102*

I

Indian pipe, *41*
Indicator, ecological, *157*
Individuals, numbers of, 179
Insect, collection at light, *181*
Insecticides, 19
Insectivorous plants, 45, 68
Invertebrates, 28, 92
  Bergman's rule for, 181
  numbers of, 180
Isothermal line, 128

J

Jerboa, *165*
Jordan's rule, 183

K

Kaibab squirrel, *139*
Kangaroo rat, 36, *37,* 88
Key-industry animals, 63
  of sea, *176*
Krakatoa, 105, 136

L

Laboratory studies, 81
 of chickens, 179
 at lake, *160, 161*
Labrador current, 149
Ladybird beetles, 18
Lake,
 biology, *160*
 Michigan, 107, *119, 120, 123*
 see Pond succession
 as *seral stage*, 161
Lava, 105, *106*
Lemming years, 183
Lichens, 41, 93, *153*
Light, 32
 collection of insects, *181*
 as limiting factor, 32
 penetration to sea, 170
 struggle for, 39
Limiting factor, 21
 light as, 32
 temperature as, 31
 substratum as, 26
Limpet, *174*
Littoral zone, 170
Littorina zone, 172, *173*
Lizard in burrow, 36
Lizard, Panama, *169*
Log community 59, *103, 104*
Long-day plants, 178
Lunar rhythms, 96

M

Mackerel, *54*
Mangrove, 23, *24*
Mantis, brown, *168*
Marine, see Ocean
Marsh, 107, 110
Maximum limits, 22
Meadow community, *65*, 79
Meadowlark, 69
Medium, 22
Merriam life zones, *151*
M*etasequoia,* 145, *146*
Mice, fighting, 55, *57*
Middle beach, 114, 119, 120
Migration of birds, 94
Mimicry, 33
 bee-mimic, *168*
Minimum conditions, 22
Mississippi River, *144*
Mistletoe, *40*
Monkey, *167*
Moon, rhythms, 96
Moose, *156*
Moss, 37

Moth,
 bee-mimic, *168*
 collected at light, 181
Mountain zonation, 150, *151*, 181
Mt. Washburn, *151*
Mt. Washington, 130
Muddy shore, *171*
Musk ox, *152*
Mutualism,
 animal-animal, 50, *51*
 artificial, 81, *82*
 plant-animal, 45
 plant-plant, 41

N

Natural history, 4
Natural selection, 67, 179
 in social evolution, 51
Neritic zone, 170
Niche, 61, 62, *65*
 basic, 66
 decomposers, 64
 dominants, 62
 food producers, 62
 herbivores, 63
 key-industry animals, 63, *176*
 numbers of, 180, 183
 parasite, 64
 scavenger, 64
 specialized, 66, *68*
 transformers, 64
Night-blooming cereus, *87*
Night-day changes, 84
Night monkey, *167*
Noctiluca, 176
Nomadism, *165*
Numbers,
 of animals, 76
 of individuals, 180
 of species, 179

O

Oak forest, 58, 117, *119, 121*
Oak-hickory forest, 85, 125
 in winter, 91
Ocean,
 biome, 170-177
 day-night, 89
 food web, 77
 gradients, 180
 see *also*, Sea
 pelagic, 170
 seasonal changes in, 95
 zonati on, 170
Ocotillo, 38
Ohio forests, 9
Opossum, *159*
Optimum temperature, 29

Overgrazing, *15, 162*
  in China, 9
Owl,
  barn, *3*
  tawny, 80

P

Paleotemperatures, 130
Panama land bridge, 138
Paramecium, 101, *102*
Parasite,
  animal, 50
  chain, 78
  of crops, 17
  niche, 64
  plant, 41
  tick, *78*
Peck-order in chickens, 55, *57*
Pelagic community, *177*
  divisi on; zone, 170
Penguin, *24, 154, 155,* 181, 182
Periodic changes, 83-97
  lunar, 96
Peripatus, *132*
Periwinkles, *173*
Petrified forest, 145, *146*
Photoperiodism, 178
Physical environment, 21-38
Pine forest, 116, *119, 121,* 148, *156*
Pioneer stage,
  definition, 100
  in dune succession, 114
  in pond   ", 108, *111*
  in protozoan, ", 101, *102*
Pitcher plant, 68
Pituitary gland, 95
Plagues of animals, 183
Plankton, *176*
  centrifuging, *160*
  plant, *70, 176*
  in rivers, 138
Plants,
  adaptati on to desert, *164*
  climate, 178
  desert, *37, 164*
  temperature regulation of, 28
  lice, see Aphid
Poikilothermal, 28, 92, 180, 181
Polar, day-night, 88
Polar bear, 128
Pollen-grain analysis, 129
Polluti on, 18
Pond community, 59
Pond successi on, 107. *111, 112*
Populati ons,
  estimating, 80
  human, 6

Porgies (scup), *177*
*P*ost climax, 124
Prairie,
  dog, *163*
  geographic distribution of, *144*
  low, 112
  meadow lark, 69, *111*
  peninsula, 132
  in pond succession *111*, 112
Predator-prey, 46
Protective resemblance, *168*
Protozoan succession, 100, *102*
Pumpkinseed, 161
Pycnogonid, *177*
Pymatuning, *160, 161*
Pyramid of numbers, *front cover, 76*

Q

Quail, *163*

R

Rabbit, *159*
  Himalayan, *32*
Raccoon, *159*
Rain, 9
Rainfall across U.S., *144*
Rain forest biome, 150
  see Tropical forest
Rain shadow, *143*
Redwoods, 142, 145, *146*
Reforestation, 2, 16
Relicts, glacial, 132
Rhythms, lunar, 96
Rocky mountains, 144
Rocky shore, *171*
Rodents, *163*
Round-tailed ground squirrel, *164*
Rule,
  Allen' s, 182
  Bergman's, 182
  Jordan's, 183
Runoff, 16

S

Sagitta, 20, *176*
Salamander, cave, *35*
Salinity, 137
Salton Sea, 140
Sand dollars, *177*
San Francisco Peaks, *151*
Sandy shore, *175*
Saprophyte, *41*
Sargassum, 23, 26
Scavenger,
  dung beetle, *68*
  hermit crab, *68*
  niche of, 64
Schooling of mackerel, *54*

Scup (porgies), *177*
Sea, *see also,* Ocean
  spider, *177*
  zonation in 170, 172, *173, 174*
Seals, *155*
Seasonal changes, 90, 91, 92, 95
Sequoia, 145, *146*
Serpent star, *177*
Sessile animals, 26
Sexual contact, 52
Shadscale desert, *151*
Shark sucker, *25*
Shore, sea, *171, 175*
Short-day plants, 178
Sierra Nevada Mts., *144*
Size of animals, 180
Size of food, 72
Sloth, *167*
Smoke tree, *164*
Snake,
  garter, *65*
  tree, *167*
Social,
  hierarchies, 55, *57*
  tolerance, 53
Societies,
  ant, *56, 163*
  human, 126, 127
  termite, 55
Soil conservation, 8
South America interchange with N.A.
  142
Spanish moss, *40*
Species, numbers of, 179
Spider, *68*
  sea, *177*
Sponge-crab, 47, *48*
Spruce forest, *153*, 156
Storm,
  dust, *11*, 13
  as natural selection, 21
Streamline form, 23, *24, 25*
Subclimax, 122, 124
Submerged vegetation, 108, *111*
Substratum, 22, 128, *25*
Successi on, ecological, 98-127
  definition, 98
  human, 126, *127*
  Krakatoa, 105
  log, 103, *104*
  lava, 105, *106*
  pond, 107, *111, 112*
  protozoan, 100, *102*
  sand dune, 113, *119-123*
  tree hole, 99
Sunstroke, 29
Supercooling, 27
Symbiosis, 41

T

Tapeworm, 50
Temperate forest,
  biome, 147
  seasonal changes, 91
Temperature,
  adaptati ons to, 179
  ancient, 130
  in desert, 87
  extremes of, 27
  as limiting factor, 31
  in ocean, 95
  optimum, 29
  and size, 180
  in tropics, 93
Temporary pond, 110, *111*
Termites, 50
Territoriality, 80
Thermometer, cricket, 29
Thomson, Wyville, ridge, 137, *138*
Tick, *78*
  bird, 66
  eating niche, 66
Tide, 96
Tissue culture mutualism, 80
Toad, *89, 161*
Tolerance, limits of, 22
Topsoil, 5, 7, 12
Transformer, niche of, 64
Tree,
  adaptations to, *169*
  frog, *44*
  growing alone, 39, *40*
  hole, *99*
  line, mountain, *151*
  snake, *167*
Tropical forest biome, *166*
Tropics, seasonal changes, 92
Tropical frog, *44*
Tube anemone, *175*
Tundra, 93, *151*
  ancient, 129
  bi ome, 148, *152, 153*
Turkey, *158*

U

Unbalance in numbers, 183
United States, *144*
Urechis caupo, 47

V

Vegetative zones, mountain, 151
Vertebras, numbers of, 183
Viewpoint, ecological, 1
Vole, 3

W

Walking-stick, *168*
Warm blooded, 29, 96, 180-182
Wasp,
  day-night, *85*
  society, 56
Water,
  cycle, *143*
  physical factor, 34
  shortage, 13
Wet-sand margin, 113
Whale,
  distribution of, 129
  food web, *73*
Wharf piling, *25*

Wind,
  and climate, *144*
  erosion, *11*
Winter, 91, 93
Wolves, 73
Wyville Thomson ridge, 137, *138*

XYZ

Yellowstone Park, *151, 156*
Yucca and yucca moth, *46*

Zonation,
  mountain, 150
  in sea, *170*
  on sea shore, *172, 173*